WAFFEN SS

WAFFEN SS

KEITH SIMPSON

GALLERY BOOKS
An imprint of W.H. Smith Publishers Inc.
112 Madison Avenue
New York, New York 10016

Published by Gallery Books
A Division of W.H. Smith Publishers Inc.
112 Madison Avenue
New York, New York 10016

Produced by
Brompton Books Corp.
15 Sherwood Place
Greenwich, CT 06830

ISBN 0-8317-2673-3

Printed in Hong Kong

10 9 8 7 6 5 4 3 2 1

CONTENTS

PAGE 1: March past of an SS band 1939.

PAGE 2: A Waffen SS recruiting poster dating from 1941. It reads: 'You can join them on your seventeenth birthday.'

LEFT: *Leibstandarte* soldiers take part in one of the victory parades celebrating the successful campaign in the West, 1940.

AN apparently elite army of nearly one million men under the command of Reichsführer SS Heinrich Himmler, the Waffen SS (Armed SS), offers many contradictions. It was condemned by the International Military Tribunal at Nuremberg as an integral part of the SS and as a criminal organization responsible for war crimes and crimes against humanity. But former Waffen SS soldiers and other apologists have claimed, in the words attributed in 1953 to West Germany's Chancellor Konrad Adenauer, that they had been *soldaten wie die anderen auch* ('just soldiers like the others'). These veteran Waffen SS men formed the *Hilfsgemeinschaft auf Gegenseitigkeit* (HIAG, or Mutual Help Association), which was primarily a welfare body but quickly became the public relations office of the Waffen SS. Senior officers such as former SS Oberstgruppenführer Paul Hausser and SS Obergruppenführer Felix Steiner defended the Waffen SS through their polemic writing.

Hausser in his book *Waffen SS im Einsatz (Waffen SS in Action)* wrote that 'he hoped the work would disperse the lies and calumnies which had formed themselves around the Waffen SS and help to give this courageous formation its place alongside the other branches of the Wehrmacht which rightly belongs to it.' Military respectability was given to the Waffen SS in postwar testimony from former senior Wehrmacht officers like Heinz Guderian who wrote that 'the longer the war went on, the less distinguishable they became from the army.'

Another claim by Waffen SS veterans was they they had belonged to the first volunteer European army united in a crusade against Communism. Felix Steiner wrote in his book *Die Freiwilligen (The Volunteers)* of 'over a half million volunteers of foreign nationality' who 'followed the dictates of their conscience,' and voluntarily left their families 'to offer their lives for a great concept' – the defense of Western Europe against the

BELOW: SS men in full ceremonial costume parade Nazi Party standards at the 1933 Nuremberg Rally.

Soviet threat through service in the Waffen SS. This opinion was taken up by former SS Standartenführer Otto 'Scarface' Skorzeny, who in 1969 wrote in the foreword to a series of books on the Waffen SS, that they 'clearly show the Waffen SS formed for the first time a European army, composed only of volunteers and united by a European ideal which was possibly premature at this time.'

During World War II the Waffen SS gained a reputation among German soldiers and civilians, as well as its opponents, of being an elite military force which displayed outstanding fighting skill both in defense and attack. But can the Waffen SS really be called a purely military unit and compared with other outstanding formations in the German Army like the *Grossdeutschland* and the *Panzer Lehr* Divisions, or with the US Rangers, the British Paratroopers or the Soviet Guards Divisions? The reputation of the Waffen SS is bound up with its direct involvement in well-documented war crimes against both soldiers and civilians, and its personal loyalty to Adolf Hitler and association with Heinrich Himmler.

Even among German civilians during the war the Waffen SS had a macabre reputation. In March 1942 Himmler's Security Service, the SD, produced a report based on a sample of public opinion which concluded: 'It may be stated that by its achievements the Waffen SS has won its place in the popular esteem.' But, 'critical voices are to be heard saying that the Waffen SS is a sort of "military watchdog." SS men are trained to be brutal and ruthless, apparently so that they can be used against other German formations if necessary.' The overall impression was that 'the Waffen SS is the most ruthless force; it takes no prisoners, but totally annihilates its enemy.' The origins of this 'ruthless force' lay in Hitler's requirement for a personal bodyguard and a force which could be used against the enemies of National Socialism.

BELOW: SA and SS men parade together at a Nazi Party Rally, 1933. The power of the SA was destined to be destroyed, along with its leadership, in the blood purge of June-July 1934.

ON 30 January 1933 Adolf Hitler was sworn in as Germany's chancellor, and on taking up residence in the Reichs Chancellery Hitler found himself protected by a detachment of the *Reichsheer*, the army, and some uniformed police. He felt insecure knowing that his personal safety depended solely upon state security forces, so on 17 March Hitler ordered Joseph 'Sepp' Dietrich to form the SS *Stabswache* (SS Headquarters Guard), Berlin, consisting of 120 selected men from Hitler's personal bodyguard in Munich. The *Reichsheer* units were eventually withdrawn leaving Hitler's security to the SS.

The origins of the SS *Stabswache* lay in Hitler's wish to maintain a small, absolutely loyal, politically reliable group of physically fit toughs, sworn to protect him with their lives if necessary. Originally formed in

forced to allow Dietrich considerable independence because Hitler allowed no one to interfere in matters of his personal security. Hitler trusted Dietrich, an *Alte Kämpfer*, a totally loyal tough with a coarse sense of humor, in fact in Hitler's own words, 'he was the original Bavarian.' Dietrich was born into a Swabian working class family in 1892 and had served in the Bavarian artillery, a stormtroop section and finally with a tank unit during World War I. Dietrich left military service in 1919 as a non-commissioned officer and joined the police. He served with the *Freikorps* in Silesia and then resigned from the police. After a series of jobs – clerk, customs official and gas station attendant – Dietrich joined the Nazi Party in 1928 and soon gained Hitler's attention by his handy use of his fists at political meetings and events.

CHAPTER 1

Special Disposal Troops

March 1923, the *Stabswache* had then consisted of two men but over the next two years it was reorganized and in 1925 was established as the *Schützstaffel* (SS, or Protection Squad). The fact that the SS evolved into something more than Hitler's bodyguard was mainly due to Heinrich Himmler who was appointed Reichsführer SS on 16 January 1929. Hitler ordered Himmler to make the SS into 'an elite troop of the Party,' smaller and more dependable than the larger and less reliable *Sturmabteilung* (SA, or Storm Section). By the time Hitler became chancellor the SS had some 52,000 members as against 300,000 in the SA, but whereas Himmler restricted further recruiting, Ernst Röhm, the SA chief of staff, allowed the SA to grow to over a million men within a few months.

Sepp Dietrich's SS *Stabswache* formed a tiny part of the SS, and the overwhelming majority of SS men were part-timers in the *Allgemeine SS* (General SS). Himmler was

With his rough, open Bavarian manner, his war record, physical strength, total loyalty and utter ruthlessness, Dietrich was just the man to run the SS *Stabswache*. In May 1933 the *Stabswache* was formed into the SS *Sonderkommando Zossen* and moved into barrack accommodation in the former military cadet college at Berlin-Lichterfelde. During September, at the Nuremberg Party Rally, Hitler named the guard battalion after himself – the *Leibstandarte Adolf Hitler*. Two months later, on the tenth anniversary of the Munich Beer Hall Putsch, the *Leibstandarte* paraded in front of the Feldherrnhalle War Memorial in Munich and took a personal oath of loyalty unto death to Hitler. So within eight months of becoming chancellor, Hitler had created a personal bodyguard quite separate from the army and the police, one armed, bound to him by an oath and available as an instrument of political terror.

Concurrent with Dietrich establishing the

LEFT: Reichsführer SS Heinrich Himmler, founder and leader of the SS.

LEFT: Sepp Dietrich, first commander of the *Leibstandarte*, wearing the Knight's Cross and Oak Leaves, Russia, January 1942.

BELOW: A poster urging Flemings to join the ranks of the SS, giving some credence to the concept of the Waffen SS being a Pan-European force.

RIGHT: A Waffen SS recruiting poster displayed in Berlin during 1940.

ABOVE: Ernst Röhm (third from right) inspecting units of the SA in 1933. A glum-looking Himmler (far left) ignores the ceremony.

LEFT: Hitler, accompanied by Himmler, inspects a *Leibstandarte* honor guard.

BELOW: SS and SA men display captured Communist flags and placards taken during a 'political action.'

SS *Stabswache*, a number of SS administrative regions created *Politische Bereitschaften* (Political Alarm Squads) which acted as personal guards for local SS leaders and as an auxiliary police force to threaten political opponents. Eventually all of Germany was covered by a network of *Politische Bereitschaften* each training armed SS men for political action on the home front. A number of SS *Sonderkommandos* were despatched to set up and operate concentration camps for political prisoners, such as at Papenburg and Dachau. Turned into SS *Wachverbände* (Guard Units), they were to form the basis for the infamous SS *Totenkopfstandarten* (Death's Head Regiments).

The *Leibstandarte*, the *Politische Bereitschaften* and the *Wachverbände* were ultimately all instruments of the Nazi Party – bodyguards, paramilitary forces and in-

ternal security units combined. They protected the Nazi Party against its external political enemies – socialists, communists and conservatives – and from opponents within the party itself. In 1934 Hitler, Goering and Himmler saw the internal threat coming from Röhm and the SA who appeared to be advocating radical policies beyond Hitler's immediate aim of consolidating his power. In particular, Röhm and the SA were challenging the *Reichswehr's* prerogative to be the 'sole bearer of arms' in Germany.

On the 30 June the SS demonstrated its unquestionable obedience to Hitler. Convinced that the SA leaders were planning a putsch, Hitler decided to eliminate his opponents in the Nazi Party. Two companies of the *Leibstandarte* moved to Bavaria and arrested Röhm and other senior SA members. SS men from the *Leibstandarte*, the *Politische Bereitschaften* and the *Totenkopfverbände* manned the firing squads in Munich, Berlin and other cities, carrying out the death sentences ordered by Hitler, Goering and Himmler. On 26 July Hitler rewarded the SS for its meritorious service during the 'Blood Purge' and elevated it to the position of being an independent organization within the Nazi Party. The *Leibstandarte*, the *Politische Bereitschaften* and the *Totenkopfverbände* had proved to Hitler that they would destroy his enemies, even if they were former political comrades. Hitler now saw them as forming the basis of a reliable internal security force which would protect the Nazi regime against political opponents in peacetime and prevent a repetition of the experience of 1918, of the political 'stab-in-the-back,' in wartime.

On 16 March 1935 Hitler announced that

ABOVE: Hitler, surrounded by his SS bodyguard, moves among the adoring crowds at one of the Nuremberg Rallies.

LEFT: 'The Asphalt Soldiers.' *Leibstandarte* parade before Hitler, Himmler and Dietrich along the Wilhelmstrasse in Berlin, 1933.

BELOW: Known as 'The Man with the Iron Heart,' SS Gruppenführer Reinhard Heydrich was appointed to lead Himmler's security services.

creation of the Luftwaffe, to many Germans the SSVT looked insignificant.

The SSVT was to take its place alongside the other repressive organs of the Nazi regime. In 1936 Himmler combined the appointments of Reichsführer SS and Chief of the German Police, and divided the police apparatus into two branches. Under the command of SS Obergruppenführer Kurt Daluege was the *Ordnungspolizei*, the uniformed Order Police, and after 1939 Himmler was to create over 100 battalions of Order Police for security duties in the occupied territories. SS Obergruppenführer Reinhard Heydrich was appointed Chief of the Security Police and the SD, combining the Gestapo, the Criminal Police and the Security Service. But Himmler saw himself and the SS as more than crude instruments of Nazi security policy: the SS was to be a racial and ideological elite which would transform German society and act as a vanguard in establishing a Nazi-dominated Europe. Himmler was a frustrated soldier. Too young to have seen active service during World War I, Himmler had served briefly as an officer cadet and then had experienced a form of pseudo-military life in the *Freikorps* and the Nazi Party. For Himmler the SSVT represented the two strands of his personality and political creed.

Himmler faced two immediate problems with the SSVT: finding suitably experienced military instructors and obtaining sufficient equipment. He was able to recruit a number of competent former professional army officers because he gave them the impression that the SSVT would have a traditional military role alongside the army, and yet be genuinely open to new talent and fresh ideas. An early recruit of this caliber was retired *Reichsheer* Lieutenant General Paul Hausser who transferred from the SA to the SS. Born in 1880, Hausser, a Brandenburger, had been a career soldier and was a qualified General Staff officer. Himmler recognized that he was just the man to provide the SSVT with the necessary military skills. In 1935 Hausser opened an SS *Junkerschule* at Brunswick, followed by a second at Bad Tölz, which would train the future officers of the SSVT. By the end of May 1935 the SSVT had 8459 men, with 2660 in the *Leibstandarte*, 759 attached to the SS *Junkerschulen*, and the remainder divided among the six battalions of the two SSVT regiments. In addition there were 1338 SS garrison troops and 2441 *Totenkopfverbände* guarding the concentration camps.

Although former members of the Waffen SS vigorously deny any connection with the *Totenkopfverbände*, in the opinion of both

Germany was re-introducing conscription, intending to form an army of 36 divisions and creating an independent air force, the Luftwaffe. Persuaded by Himmler he also announced the formation of the SS *Verfügungstruppe* – (SSVT, or Special Disposal Troops). The SSVT was to be formed from the *Politische Bereitschaften* and the *Leibstandarte* and consisting of fully motorized formations, was intended to serve as a nucleus for an eventual SS division. The German Army was deeply suspicious of the SSVT seeing it as challenge to the army's monopoly of military power. However, despite Himmler's ambitions, Hitler had no intention of turning the SSVT into a rival army. The SSVT was an internal security force whose disposal in the event of war would be decided by him personally. With the massive expansion of the army and the

Hitler and Himmler the SSVT and the *Toten-kopfverbände* were equally important in combating the enemies of Nazi Germany. On 1 April 1936 it was stated that the SSVT and the *Totenkopfverbände* were 'organiza-tions in the service of the state' and placed on the police budget of the Ministry of the Interior.

In October 1936 Himmler appointed Hausser inspector of the SSVT to supervise its military training. Hausser amalgamated the scattered battalions of the SSVT into the SS Regiment *Deutschland* under the com-mand of SS Standartenführer Felix Steiner in Munich, and SS Regiment *Germania* under SS Standartenführer Karl Demelhuber in Hamburg. Nominally, Hausser had the authority to inspect and oversee the training of the *Leibstandarte*, but for many months he was refused access to the unit by Dietrich. No greater contrast could be found in the SS than that between Hausser, the austere, conservative former professional officer and retired general, and Dietrich, the

flamboyant, beer-swilling, former non-com-missioned officer and Nazi thug. Dietrich jealously guarded his independent com-mand and through his close personal re-lationship with Hitler was able to frustrate Himmler and Hausser. Eventually, Dietrich realized that the members of the *Leibstan-darte* were acquiring a reputation as 'Asphalt Soldiers,' a nickname given to them by the rest of the SSVT on account of their frequent ceremonial duties. Dietrich relented, Hausser gained access and the *Leibstandarte* began to train for war.

The SSVT had no problems in attracting sufficient numbers of recruits, despite the very stringent requirements for what was seen as an elite force. Service in the SSVT was voluntary and counted toward fulfilling military service. Long periods of enlistment were required from potential recruits – four years for soldiers, 12 for NCOs and 25 for officers. A very high standard of physical fit-ness was required from the applicants who had to be aged between 17 and 22. A recruit

TOP: Hitler, Hess, Himmler and Dietrich review the *Leibstandarte.*

ABOVE: SS Obergruppenführer Felix Steiner.

RIGHT: The massed ranks of SS in the Luitpold, 1936.

had to be a minimum of 5ft 10 inches tall, later 6ft 5 inches for the *Leibstandarte*. Himmler insisted that every SS recruit had to be 'of well-proportioned build; for instance there be no disproportion between the lower leg and the thigh, or between the legs and the body; otherwise an exceptional bodily effort is required to carry out long marches.' In 1943 Himmler was to claim that 'Until 1936 we did not accept a man in the *Leibstandarte* or the *Verfügungstruppe* if he had even one filled tooth.' Himmler's emphasis on physical fitness was ironic given his gawky appearance, short-sight, stomach cramps and inability to pass his own SS physical proficiency test.

But Himmler went beyond simple physical criteria, demanding that the potential recruit fulfill the racial requirements of an Aryan pedigree and appearance. Every SS man had to produce his ancestry back to 1800, and officers back to 1750. Like a butterfly collector, Himmler would examine photographs of recruits with a magnifying glass seeking evidence of Jewish or Slav blood.

Who joined the SSVT and why? Initially, the officer and NCO cadre came from former members of the army and police, or from those who had served with the *Freikorps*. The original 120 SS men of the *Stabswache* had been young political toughs employed as bodyguards and bouncers by the Nazi Party. The overwhelming majority of volunteers in the SSVT were working class and mainly of peasant stock. In 1938 less than two percent of army officers were of peasant stock as against 90 percent in the SSVT. Some 49 percent of army officers came from military families; only five percent in the SSVT. Reflecting their working class background 40 percent of SSVT officer candidates in 1938 had only an elementary school education.

The SSVT was a vehicle for social and professional mobility for young, working class men with basic education and no military background. For men like Fritz Witt, Otto Kumm and Kurt Meyer, it offered a career in a crack Nazi unit. Nazi propaganda promoted the SSVT, especially the *Leibstandarte*. Smart uniforms, ceremonial duty, Hitler's personal approval, a reputation for physical fitness and a form of military service all combined to make the SSVT seem very attractive to German youth.

Basic training in the SSVT was the same for all recruits – officer candidates served two years in the ranks before being sent to an SS *Junkerschule*. Recruits endured traditional German army training but a greater emphasis was placed on field craft, hand-

ABOVE: SSVT emphasis on physical fitness helped to create an ideological elite, one fit to combat the rigors of prolonged campaigning.

LEFT: SSVT recruits, dressed in fatigues, pose for the camera after collecting their midday meal.

ling weapons and sport, with the addition of ideological lectures. Once a recruit had passed the initial training he was eligible to take the SS oath. SSVT candidates took this oath separately from other SS members at 2200 hours on 9 November before the Feldherrnhalle in Munich in Hitler's presence. This very emotive ceremony was designed to bind the SSVT man to the person of Adolf Hitler:

'I swear to thee Adolf Hitler
As Führer and Chancellor of the
    German Reich
Loyalty and bravery.
I vow to thee and to the superiors
    whom thou shalt appoint
Obedience unto death
So help me God.'

ABOVE: SS Standartenführer Otto Kumm, an officer of humble origins, rose through the ranks of the SS and ended in command of the *Prinz Eugen* Division.

RIGHT: Members of the SSVT examine a target on a rifle range.

BELOW RIGHT: Hitler, accompanied by Dietrich, inspects a *Leibstandarte* barrack room.

Following a year at an SS infantry or cavalry school the SSVT candidate returned to Munich to swear another oath binding himself to Himmler's marriage laws. For the successful officer candidate there was the presentation of the coveted SS dagger and SS signet ring.

The SSVT soldier experienced a radically different form of political indoctrination and military training from his contemporaries in the army. During his service in the SSVT he was never allowed to forget that he was a

ABOVE LEFT: An SS baptism takes place against a backdrop of Nazi iconography.

ABOVE: A nurse places the children of SS parents out on the veranda of a maternity home.

LEFT: SS officers receive their much-prized ceremonial daggers.

ABOVE: The trappings of the quasi-religious ceremony created for members of the SS.

tues. SSVT candidates were encouraged to abandon their traditional Christian beliefs, leave the Church, and adopt the Nazi form of atheism called *'Gottgläubigkeit'* or ('Belief in God'). Membership of the SSVT reinforced a patriotic belief in Germany. The war would test and harden the Nazi faith which had been hammered in during basic training and through service in a special force.

Once in the SSVT the aim of the officers was to create 'a supple, adaptable soldier, athletic of bearing, capable of more than average endurance on the march and in combat.' He was to be 'as much at home on the battlefield as on the athletics field.' The leading advocate of SSVT training was Standartenführer Felix Steiner who was a military radical. A former professional army officer, Steiner had rebelled against the conservative values and doctrine of the *Reichsheer*. In World War I he had led a stormtroop company and seen the potential of combat groups recruited from the best, most physically fit soldiers, with none of the usual formal distinctions of class and rank between officers and men. The new soldier would be trained to fight at close quarters with automatic weapons and light, mobile artillery. Steiner was convinced that victory in the future would be won by top-notch fighting groups rather than large conscript armies. Finding that his views found little favor in the *Reichsheer*, Steiner was a natural candidate for the SSVT. His energy, enthusiasm, ideas, iconoclastic methods and his conflict with the traditionalists in the army all combined to appeal to Himmler.

Steiner began to put his theories into practice and found that others in the SSVT, including Dietrich, were in many respects already ahead of him. Steiner relegated formal drill to the bottom of his list of military priorities. Sport and athletics were stressed so that after training the SSVT recruit could cover two miles in 20 minutes while in full battle order of 60lbs. More time was spent in the field, on the ranges, and in the classroom learning the theory of tactics than was practiced in the army. Steiner introduced into the SSVT submachine guns and more handgrenades than were available in the army. He wanted to organize the SSVT into mobile battlegroups that could quickly come to grips with the enemy and then adapt with mobility and flexibility to the rapidly changing circumstances of the modern battlefield.

Training exercises were made as realistic as possible, the conditions of combat being simulated with live ammunition and artillery fire. Casualties were inevitable but

member of an elite Nazi cadre. Political and ideological indoctrination was given parity with other phases of training – and during the war this continued. Until 1936 it was the role of specially selected SS instructors, but then Himmler decided to restrict their position to pure supervision, giving company officers the responsibility for their own men.

As volunteers all SSVT personnel knew that they were joining a Nazi select, and that their selection had been determined by Nazi ideology. Although the majority of SSVT recruits were unable to discuss the finer points of Nazi philosophy, they accepted its basic tenets. Life was a racial struggle between the Aryans and inferior races such as Jews and Slavs; Hitler and the Nazi Party had saved Germany from the Jews and Communism. Life was a struggle in which the stronger would dominate the weaker, and unquestioning obedience to the leader, mental and physical hardness toward oneself and others were noble vir-

accepted as a price to pay for military efficiency. The end product of the training was to produce a fighter rather than a soldier, someone who, when working with his comrades would, in Steiner's words, 'by blows of lightning rapidity, split the enemy into fragments and then destroy the dislocated remnants.' Dietrich, Steiner, Keppler, Bittrich and other unit commanders were keen to advocate a genuine sense of camaraderie. They were convinced that a close relationship formed in the barracks, in the mess, on the playing fields and on military exercises would pay dividends in war. In contrast to the army, officers and NCOs frequently ate and drank with their men. By 1939 the overwhelming majority of junior officers and NCOs had served in the ranks. In the *Leibstandarte* strict forms of address were discounted, and it was not unusual for a soldier to address an officer, 'May I speak to you as a man of the *Leibstandarte*?'

The uniform of the SSVT, whether parade, barrack dress or field service, helped to accentuate a sense of eliteness. Before the war, for ceremonial, guard duties and 'walking out,' the black SS uniform was the standard dress. The SS death's head on the cap badge was adapted from *Kaiserheer* cavalry insignia. Members of the *Leibstandarte*, and later members of other SSVT formations and Waffen SS regiments and divisions, wore a distinctive cuff armband bearing the name of the unit. On the belt buckle instead of the army's '*Gott mit uns*' was emblazoned '*Meine ehre heisst treue,*' ('Loyalty is my Honor'). Other distinctive insignia were the SS version of the eagle worn on the cap and on the left upper sleeve, and the collar patches and rank insignia. A profusion of insignia in wartime guaranteed distinctions between newly formed units. By 1937 the SSVT was wearing a field-gray uniform similar to the army for military training, but a year later Steiner introduced, on a limited scale, camouflage combat clothing in the form of a smock and helmet-cover. Worn over the service dress, until 1942 it provided the easiest way of identifying Waffen SS units on active service. When first worn by the SSVT, camouflage jackets earned the derision of the army which accused the SSVT of 'running around like tree frogs; real soldiers wear field-gray.'

Political and ideological indoctrination, realistic training for war, camaraderie and the symbols of elite status characterized the SSVT. The final significant factor in developing the unique nature of the SSVT was the atmosphere of 'heroic realism' which permeated its ranks. This owed its philosophical origins to Friedrich Nietzsche who

LEFT: Recruits swearing the SS oath.

BELOW LEFT: Hitler inspecting an SSVT honor guard in Munich.

ABOVE: Ernst Jünger.

BELOW: Hitler salutes Reinhard Heydrich's coffin in June 1942.

stressed that in order to deal with a chaotic universe, man had to abandon himself to struggle and combat for its own sake. This was to be the experience of the young *Frontkämpfer* during World War I. Their most articulate spokesman was Ernst Jünger, who as a young storm troop officer had won the *Pour le Merité*. After the war in a series of powerful and influential books and essays – *In the Storm of Steel, Fire and Blood* and *Combat as Inner Experience* – Jünger portrayed battles as 'a magic delight.' It was a 'magnificent show of destruction' and 'a splendid miracle: a transformation of the bourgeois into the adventurer.' Jünger also wrote of a 'new man, the storm soldier, the elite of central Europe. A completely new race, cunning, strong and packed with purpose .... battle proven, merciless both to himself and others.' Jünger might have been describing the future graduates of the SS *Junkerschulen*. This concept of fighting for fighting's sake turned the traditional

soldierly concept into one of pure belligerence. It gave both the leaders and the led a distinctive bravado and *Härte* (harshness or severity) toward oneself, one's comrades and, of course, toward the enemies of the Führer, the SS and the Germanic race.

This atmosphere of heroic realism reached its ultimate apotheosis in the ranks of the junior officers of the SSVT. While at the SS *Junkerschulen* and then at regimental duty SS officers were inculcated with a fatalistic and passionate enthusiasm for combat which went far beyond the normal self-sacrifice that might be expected from a soldier. During the war the idol symbolizing SS heroic realism was SS Obergruppenführer Reinhard Heydrich, Chief of the Security Police and the SD, who, in Nazi mythology was struck down on the field of battle in 1942. In fact Heydrich was acting Reich Protector in Bohemia and Moravia when he was assassinated by Czech agents. The ethos of heroic realism helps to

explain the particularly heavy casualties among Waffen SS units, and the determination and hardness of the survivors who, as relatively young men, were to command the battalions, regiments and even divisions of the greatly expanded wartime Waffen SS.

For Hitler and Himmler the primary role of the SSVT was to maintain internal security within Germany. In the event of war Hitler wanted an obedient, disciplined, well-armed and totally ruthless force of Nazis capable of crushing any internal unrest. Hitler and Himmler knew that the SSVT would find it difficult to carry out its duty of internal security in war and maintain the respect of the German people if it had not undertaken combat at the front. As Himmler said to senior SS officers in 1938, by sacrificing its 'blood at the front,' the SSVT would retain 'the right to shoot malingerers and cowards on the home front.' Ultimately Himmler saw the SSVT as forming the core of a *Staatsschutzkorps*, a State Protection Corps, which would guard a Germanic empire in Europe.

The army still viewed the SSVT as a potential rival to its miltary monopoly. Professional rivalry, contempt for an organization led by Nazi thugs, former NCOs and policemen, combined with a determination to restrict the size of the SSVT, led the army

LEFT: *Totenkopf* guards at Dachau Concentration Camp oversee the distribution of food to 'political undesirables.'

BELOW LEFT: Himmler and his entourage inspect Dachau Concentration Camp in 1935.

BELOW: Hitler pictured entering the Sudetenland in triumph, 1938.

to attempt to control the allocation of manpower, weapons and equipment. Disagreements between the army and the SSVT manifested themselves in disputes over training areas and in street brawls between soldiers and SS men. Hitler agreed with the army to restrict the growth of the SSVT, to refuse it artillery, to forbid it to publish recruiting notices in the press and to allow the army the right to inspect SSVT units.

In the winter of 1938 Himmler seized the opportunity to persuade Hitler to lift his restrictions on the role and size of the SSVT. As a consequence of SS evidence, Field Marshal Blomberg, the Minister of War, resigned when it was discovered that his wife was a former prostitute, and General von Fritsch was neutralized after the SS alleged that he was a homosexual. Hitler decided to bring the Wehrmacht more directly under his control and retire senior officers who opposed his policies while promoting those who supported them. A new Armed Forces High Command, OKW, was created under General Keitel. On 11 March 1938 German troops marched into Austria with a motorized battalion of the *Leibstandarte* in the lead formations. Units of the SS Security Police and the *Totenkopfverbände* ruthlessly suppressed political opponents.

On 17 August Hitler issued a secret Führer Decree delineating the common tasks of the SS and the Wehrmacht: 'For special internal political tasks . . . or for use within the wartime army in the event of

mobilization,' the SSVT, the SS *Junkerschulen*, the SS *Totenkopfverbände* and its reserve units were to be organized, trained and armed as military formations. In peacetime Himmler would have the responsibility for running these units and would carry out such internal security tasks as Hitler gave him. There would be no connection with the army in peacetime, but it would provide the necessary weapons and equipment. Hitler made it clear that the SSVT was 'neither a part of the Wehrmacht nor a part of the police. It is a standing armed unit exclusively at my disposal.'

The premobilization strength of the SSVT included the motorized *Leibstandarte*, three infantry regiments, and motorcycle and engineer battalions. In the event of war Hitler would personally decide how to deploy the SSVT, either at the front under the command of the army or, depending on the internal political situation, under Himmler on the home front. Under the Führer Decree of August 1938 Himmler was also authorized to provide reinforcements for the SSVT from members of the SS *Totenkopfverbände* who would 'meet the ideological and political spirit' of the unit. After the Führer Decree of 18 May 1939, Himmler could increase the strength of the *Totenkopfverbände* by another 40,000 men drawn from the *Allgemeine* SS as *Polizeiverstärkung* (police reinforcements) in the event of mobilization. These police reinforcements would provide Himmler with the manpower for more than a dozen SS *Totenkopfstandarten* (Death's Head Regiments) which could be deployed for 'special duties' in the occupied territories.

The first real opportunity for the SSVT to be deployed in a military role other than as a token force was during the occupation of the Sudetenland. The entire SSVT and four battalions of SS *Totenkopf* were placed under the command of the army. Operationally a success, the army nevertheless attempted to exclude the SSVT from the congratulatory Order of the Day. Units of the SSVT, formed as mobile assault units, were integrated with army panzer divisions that occupied the rump of Czechoslovakia in March 1939. That summer, after observing the SSVT carry out maneuvers under fire, Hitler finally gave orders for the SSVT to be established as a division with its own artillery. The crisis with Poland in August meant the postponement of this plan. Individual SSVT units were organized into regimental combat groups and attached to the army divisions preparing to invade Poland. On 1 September 1939 the SSVT went to war serving alongside the German Army.

THE German attack on Poland on 1 September 1939 was planned as a short, decisive campaign and not the prelude to a world war. Hitler, seeing himself as the spiritual heir of Frederick the Great, referred to it as his 'First Silesian Campaign.' Major units of the SSVT were moved to East Prussia, where the SS Regiment *Deutschland*, the new SS artillery regiment, the SS reconnaissance battalion and an army panzer regiment formed a brigade under Major General Werner Kempf. Hausser moved from his post as Inspector-General of the SSVT to serve as a liaison officer with Kempf. SS Regiment *Germania* was attached to the Fourteenth Army in East Prussia, while the *Leibstandarte* formed part of the Tenth Army in Silesia. Only the SS Regiment *Der Führer*, formed after the *Anschluss* with Austria in 1938, failed to see action in

darte, won the thanks of a grateful Führer and received generous publicity concerning its actions. But the campaign exacerbated the frictions between the army and the SSVT. Senior army officers criticized the SSVT's poor tactical handling and its proportionally higher casualties. SSVT officers countered this by pointing out that their units had been attached to army formations without integral artillery or support units and had been ordered to carry out difficult if not impossible tasks. This disagreement formed part of the wider dispute between the SS and the army concerning the 'special police tasks' of the SS behind the front in Poland and the establishment of German occupation. Behind the army and the SSVT Himmler had deployed mobile groups of Security Police and SD, Order Police battalions and SS *Totenkopf* Regiments, whose

CHAPTER 2

Blitzkrieg Campaigns 1939-42

Poland, being relegated to manning a section of the Westwall.

During September the SSVT, along with the German Army, carried out what became seen as the classic Blitzkrieg campaign of deep penetration, encirclement and destruction of the Polish Army. But at a tactical level the Germans discovered that the Poles were courageous and tenacious. During one fierce engagement the *Leibstandarte* found the Poles were skillful fighters: 'The Poles are devilish cunning . . . We had a mission yesterday and had to clear a group of them from a field of maize. We thought they were a handful of stragglers and that a quick sweep would finish our mission . . . They had dugouts with crops growing on the roof and were almost invisible and hard to detect. We . . . had to stalk them like characters from a Karl May Wild West novel. When we found a dugout we blew it up with bundles of grenades.'

The SSVT, and in particular the *Leibstan-*

task was to arrest, and in many cases shoot, Jews and Polish political and nationalist figures. Some army commanders were disgusted and appalled by the unrestrained brutality shown by the SS and police. The army was only too willing to hand over the administration of the occupied Polish territories to the Nazi Party and the SS. Over the winter of 1939-40 Himmler's SS and police carried out a massive population resettlement of Jews and Poles, along with thousands of executions.

With the conclusion of the Polish campaign and the beginning of the Phony War in the West, Hitler pressed ahead with the expansion of the Wehrmacht, also agreeing to the formation of three SSVT divisions and the expansion of the *Leibstandarte* to what was effectively a motorized regiment. The SSVT Division, commanded by Hausser, was formed from the three SS Regiments and support units. But the army's control of SSVT recruitment meant that Himmler had

LEFT: The early stages of Operation Barbarossa – Waffen SS troops attached to *Das Reich,* part of Guderian's Second Panzer Group, watch a burning Russian village, June 1941.

ABOVE LEFT: Led by an SdKfz 232 armored car, *Leibstandarte* men advance through a burning Polish town, September 1939.

LEFT: 'Heimwehr' troops advance during operations against the 'Polish corridor.'

ABOVE: Order Police arrest an elderly Jew in Poland, the beginning of Hitler's 'New Order' in Eastern Europe.

RIGHT: Hitler congratulates Dietrich on the *Leibstandarte*'s role in Poland. Max Wünsche looks on (center).

to consider other sources of manpower. His problems were solved by SS Brigadeführer Gottlob Berger, the head of the SS Recruiting Office. Berger, a 43-year-old Swabian, was an extrovert figure, serving as a storm-troop officer in World War I then with the *Freikorps*, before becoming a teacher. His time in the Nazi Party, SA and SS had been characterized by squabbling and public displays of emotion, but Berger used his charm, cunning, administrative skills and obsequious loyalty to influence Himmler who was impressed by his tenacious fight for the interests and rights of the SS.

Berger discovered two sources of manpower available to Himmler over which the Wehrmacht had no control. Members of the *Totenkopfverbände* and SS *Totenkopf* reinforcements, as well as the Order Police, were all exempt from service in the Wehrmacht. Himmler could reinforce the SSVT Division with men from the SS *Totenkopf* Regiments, create the *Totenkopf* Division from the SS *Totenkopfverbände* – concentration camp guards – and from the SS *Totenkopf* Regiments, and transfer 16,000 Order Police to form the *Polizei* Division. Replacements for the SS *Totenkopfverbände*, *Totenkopf* Regiments and Order Police were found by drafting in men from the *Allgemeine* SS and police reservists. Within three months the SSVT, or the Waffen SS as it was now called, had expanded from 25,000 to nearly 100,000 men. Such a massive expansion created considerable problems in terms of recruitment, training and the acquisition of sufficient weapons and equipment. Himmler was fortunate that Hitler was forced to postpone the campaign in the West from the autumn of 1939 until the spring of 1940.

During the winter of 1939-40 Berger undertook interminable negotiations with the army over questions of Waffen SS recruitment, replacement formations and the supply of equipment. Plenty of young Germans were volunteering to join the Waffen SS but the army refused to release them. The army was rigorously keeping to the OKW agreement on the apportionment of manpower which fixed the army at 66 percent, the navy at nine percent and the air force at 25 percent, with the Waffen SS getting a small quota of the army's manpower. Himmler gained a significant advantage in January 1940 when the army granted him complete authority over the *Ersatz* formations of the Waffen SS.

Each field unit of the Waffen SS had an *Ersatz* (replacement) battalion, in which new recruits were trained, and over the next two years Himmler was to establish dozens of such battalions throughout Europe where they could train recruits beyond the army's control. These battalions provided Himmler with an additional source of manpower for 'special police duties.' In March 1940 an OKW decree recognized that service in the Waffen SS applied to the SSVT, *Totenkopf* and *Polizei* Divisions and the *Leibstandarte*, the SS *Junkerschulen*, and all their replacement and training units. Significantly, the SS *Totenkopfverbände* were made part of the Waffen SS by this decree. The SSVT, *Leibstandarte* and *Totenkopf* were all designated as motorized at a time when the majority of army divisions were without mechanical transport. But all the Waffen SS units had to compete with the army for the limited number of modern weapons and vehicles that were available.

One consequence of the rapid expansion of the Waffen SS was a dilution of the strict standards that Himmler had established. The work of Hausser and Steiner to establish the military credibility of the Waffen SS appeared to be undermined by forming a division recruited from concentration camp guards. Sixteen thousand Order Police did not meet the requirements of the Waffen SS and were not considered SS men until 1942. They continued to wear their police uniforms but with an eagle on the left upper sleeve. Prewar standards for the volunteers entering the SSVT Division and the *Leibstandarte* were maintained.

With the establishment of the *Totenkopf* it was difficult for the military professionals in the Waffen SS to maintain the separation of functions between the combat and the police elements of the SS, particularly with Himmler's appointment of Theodor Eicke, the Inspector of Concentration Camps and Commander of SS Guard Formations, as divisional commander. Born in Alsace in 1892, Eicke had served as an army paymaster during World War I. He left the army after the war and soon acquired a deep hatred for the Weimar Republic, Jews, communists and the churches. Joining the Nazi party and the SS his dynamic energy, administrative skills, ruthlessness, brutality and absolute commitment to National Socialism brought him to Himmler's attention. A violent character who was always quarreling with other Nazis, Eicke was an unbalanced figure, but Himmler appreciated his peculiar talents and in 1933 appointed him commandant of Dachau concentration camp. Eicke remodeled the camp and established a code of conduct for the SS guards which was to form the basis for that of the *Totenkopf* Division. SS guards were conditioned into blind and absolute obedience to all superior

ABOVE: Theodore Eicke (fourth from right) points out details at Dachau Concentration Camp to visiting Nazi officials, including Reichsorganisationleiter Robert Ley (fourth from left). Eicke later took charge of the *Totenkopf* Division.

orders and into treating each prisoner with fanatical hatred as an enemy of the state. Eicke was a living example of Nazi harshness and his total loyalty had been demonstrated in 1934 when he had shot Ernst Röhm.

After his appointment as Inspector of Concentration Camps and Commander of SS Guard Formations, Eicke established a uniform concentration camp system throughout Germany and organized the *Totenkopfverbände* into a ruthless and obedient force. His recruits were the unemployed, uneducated yokels, social misfits and former soldiers and policemen. Eicke told his *Totenkopf* guards that they were members of a racial elite who were privileged to watch over enemies of the state. Eicke taught his men an unremitting hatred for everything and everyone that wasn't Nazi. He was completely cruel and insensitive to human suffering, hard toward his own men and their prisoners, yet a strangely charismatic leader, known to his men as 'Papa' Eicke.

When in October 1939 Eicke received his orders to establish the *Totenkopf* Division he took with him his brutal ethos and his concentration camp guards. In five months

Eicke transformed the *Totenkopf* from an assortment of poorly trained units with few weapons and equipment into a well-trained, fully motorized division. Eicke hated the army and had a low opinion of other Waffen SS units. Despite his almost total lack of military experience he set himself the task of acquiring the necessary skills, but never grasped the theory of armored warfare or the details of operational planning. But like Sepp Dietrich he adapted the army's operational doctrine of speed and surprise by concentrating all available mass and firepower in the forward units, until by reckless, zealous pursuit of the attack the enemy surrendered or was annihilated. Initially, the army viewed Eicke's division with disdain because of its unsavoury antecedents and because they suspected that it would never be an effective force. However, by the spring of 1940 the army was grudgingly impressed by the professional skills of the *Totenkopf*. Eicke became a competent divisional commander and would leave his distinctive imprint on the Waffen SS.

By April 1940 the Waffen SS was ready to participate in Operation *Gelb*, the German offensive in the West. Although numerically the Waffen SS formed an insignificant part of the 136 German divisions in the West, all of them except *Polizei* were motorized, and the troops would be fighting in Waffen SS divisions. The *Leibstandarte* and the 3rd Regiment of the SSVT Division were part of the Eighteenth Army and were to assist in the seizure of rail and road bridges on the Dutch border. The remainder of the SSVT formed part of the follow-on forces which would move into Holland. The *Totenkopf* was in army reserve near Kassel, while the less well equipped *Polizei* formed part of the garrison of the Westwall.

The *Leibstandarte*, along with the 227th Division, was to seize the bridges over the Ijssel River on the German axis of advance. The Dutch Army had made preparations to defend, and if necessary, to destroy the bridges, but its forces were poorly trained and equipped with obsolete weapons. The German attack on 10 May surprised the Dutch, and a detachment of the *Leibstandarte* seized the bridge at De Poppe, opening the way for the rest of the unit and the 227th Division. Encountering little opposition the *Leibstandarte* covered 50 miles in six hours. On the 11th the *Leibstandarte* crossed the Ijssel and SS Obersturmführer Kraas led a force 40 miles into the Dutch interior. Transferred to the 9th Panzer Division, the *Leibstandarte* raced west from Kleve toward Rotterdam. On the next day, after a heavy Luftwaffe air raid, the *Leibstandarte* passed through Rotterdam. A muddle over the surrender of some Dutch troops resulted in General Kurt Student being seriously wounded by gunfire from *Leibstandarte* soldiers. The Dutch capitulated the same day and Hitler ordered the

TOP: Grim-faced Waffen SS reconnaissance troops take a break in a French village street, May 1940.

ABOVE: A machine-gun team of the second-grade *Polizei* Division mans a strongpoint. The weapon is the Czech-built ZBvz 30.

LEFT: A river assault crossing during the fighting in Belgium, 1940.

Leibstandarte and the 9th Panzer Division to undertake a triumphal drive through Dutch towns to impress the civilian population. The Leibstandarte then drove to the Belgian border where it joined the SSVT Division.

The SSVT had seized a bridge over the Maas on 11 May and then had helped to intercept the French forces moving toward Breda. By the 14th the French had been forced to withdraw to Antwerp and the SSVT attacked the Dutch forces retiring to Zeeland. Hausser led a mixed Waffen SS and army force which captured the Dutch port of Vlissingen, forcing the French to evacuate their surviving allied troops by sea.

Until 16 May Eicke had been driven to frustration as the Totenkopf remained in reserve, then orders came for the Totenkopf to move west to join General Hoth's XV Panzer

ABOVE: The 3.7cm antitank gun was nicknamed the 'door-knocker' because it was ineffective against medium or heavy tanks of 1940 vintage.

RIGHT: Men of the SS Standarte Germania in France. Note the Luger and mortar legs carried by the man on the left.

Corps which was moving through the Allied positions. It took the *Totenkopf* nearly 48 hours, moving through massive traffic jams, to reach Hoth. On the 20th the *Totenkopf* attacked across the Sambre River toward Le Cateau and Cambrai. It was a vicious day of house-to-house fighting against French Moroccan troops and seeing off counter-attacks by French tanks.

By the following day General Guderian's XIX Corps had reached the Channel coast at Abbeville, thus driving a wedge between the French, British and Belgian forces to the north and the main part of the French Army to the south. The *Leibstandarte* and the SSVT had motored south to the Valenciennes area where they had helped to protect the northern flank of the German penetration. Meanwhile, the *Totenkopf* and General Rommel's 7th Panzer Division were preparing to cross the Scarpe River and advance northwest of Arras. As the *Totenkopf* and 7th Panzer deployed on 21 May they ran into a British counterattack aimed at breaking through the German corridor. Seventy-four British tanks and two battalions of infantry gave the *Totenkopf* and 7th Panzer a nasty shock. Fortunately for Eicke

the brunt of the British attack hit his anti-tank battalion, but their 37mm antitank guns were unable to penetrate the thick armor of the British tanks. SS infantry attacked the tanks with smallarms and handgrenades but failed to stop them. Men from the *Totenkopf* supply column fled in panic, but Eicke ordered his divisional artillery to fire over open sights and the timely arrival of some Stukas drove off the British tanks. *Totenkopf* lost 39 dead, 66 wounded and two missing in this engagement; Rommel's division suffered a similar experience.

The Allied forces in Flanders were trapped in an area against the Channel coast. To the south units of the BEF were holding positions along a canal line. The Germans redeployed and by the 24th the *Leibstandarte* was ready to cross the canal at Watten west of St Omer, while the SSVT and the *Totenkopf* were deployed farther to the east. The *Leibstandarte*, SSVT and the *Totenkopf* had established bridgeheads over the canal when Hitler issued his order to halt. He had decided to let the Luftwaffe destroy the British and French forces and save further wear and tear on his panzers. In fact Dietrich had disobeyed the halt order to

ABOVE: A makeshift stretcher used by the *Polizei* Division. The unit took part in the final battles against the French.

ABOVE: Soldiers from the SS *Germania* eat a hurried meal.

cross the canal and establish his bridge-head. When asked by Guderian why he had ignored Hitler's order, he replied that his men had seized the dominant height of the Watten because the enemy had been able to 'look right down the throat of anyone on the other bank.' Eicke was ordered to with-draw, and in doing so in the face of heavy British fire lost 42 killed, 121 wounded and five missing.

During the 48 hours before Hitler re-scinded his halt order the British consoli-dated their defensive position along the canal. With the failure of the Luftwaffe to destroy the Allies the German advance re-sumed. The British were determined to delay the Germans to allow the bulk of their troops and French forces to be evacuated by sea. The German attack included five panzer divisions, one motorized infantry division, the *Grossdeutschland* Regiment, the *Leibstandarte*, the SSVT and the *Toten-kopf*. After successfully bridging the canal on the morning of the 28th the *Leibstan-darte* was ordered to clear the village of Wormhoudt south of Dunkirk, which was defended by two British battalions. The *Leibstandarte* could only make slow pro-gress. Dietrich went forward by car to see for himself and ran into an ambush which forced him and SS Obersturmführer Max Wünsche to take cover in a ditch. As Wünsche described it; 'Fortunately, the ditch was deep, but they literally shot the edge of the ditch away with machine guns. We tried to escape to the rear. That was un-successful as a path from a field joined the road and we would have to go through a drainage pipe. Meanwhile, our car had been shot out of action; the petrol tanks leaked and a tracer round set light to the lot. I finally tried to go through the drainage pipe, but got stuck in it. I lost consciousness through what happened next.'

In fact Dietrich and Wünsche lay in the ditch for nearly five hours covered in mud to protect them from the burning petrol leak-ing from their car. Several unsuccessful attempts were made by Waffen SS and army units to rescue them and it was only after the capture of Esquebeck that they were actually extricated. Finally at the end of a frustrating day *Leibstandarte* cleared Wormhoudt. It had suffered several casual-ties, including the death of a battalion com-mander, and for most of the day there was

no news whether Dietrich was alive or dead. The majority of British prisoners were well treated, but one group of over 80 was murdered by SS soldiers of the *Leibstandarte*. This incident was probably due to the frustrations of a hard day's fighting, but clearly revealed the darker and brutal side of the Waffen SS.

The SSVT had an equally frustrating crossing and breakout from the canal line. The division had to advance through the Forêt de Nieppe. Heavy casualties among officers prompted an urgent request for replacements. The SS Regiment *Deutschland* under Steiner was attached to the 3rd Panzer Division and on 27 May advanced against a British position between Merville and Estaires on the Lys Canal. Steiner put a bridgehead over the canal and had begun to construct a crossing when 20 British tanks broke into his position. Without antitank guns and unable to get vehicles over the canal, the SS were forced to defend themselves with smallarms. Steiner watched one young SS officer set an example to his men by attempting to destroy a tank with grenades before he was crushed to death. Only the arrival of some *Totenkopf* antitank guns saved the situation.

To Eicke's fury he was forced to re-establish his bridgehead before capturing Bethune. But like the *Leibstandarte*, the *Totenkopf* found the British troops defending Bethune were determined to gain as much time as possible for the evacuation. One ineptly led SS battalion got into severe difficulties and appeared to be on the point of disin-

tegration. The commander of the battalion sent to relieve it was killed and during the day the *Totenkopf* suffered heavy casualties in the bitter fighting. At the village of Le Paradis over 100 British soldiers surrendered after running out of ammunition. They were then murdered on the orders of SS Obersturmführer Fritz Knöchlein whose company had been badly mauled in the day's fighting. During the next 48 hours the *Totenkopf* failed to break through the British rearguard and came under sustained heavy artillery fire.

With Allied troops inside the Dunkirk perimeter, the *Leibstandarte, SSVT* and the *Totenkopf* were ordered south to refit for Operation *Rot*, the forthcoming offensive against the French Army south of the Somme. Replacements arrived from Germany, and as officer casualties had been so heavy cadets were sent straight to the front from the SS *Junkerschulen*. The German plan was to launch a series of offensives from west to east with 140 divisions against 65 French divisions from 6 June. All the Waffen SS divisions, including *Polizei*, took part in the attack. Once the initial French positions had been breached, the offensive became more of a pursuit than a battle. Paris was declared an open city on the 11th and the French had difficulty in stabilizing their front. But in the days before the Armistice on the 22nd the Waffen SS found the French fighting with new determination.

The six-week campaign in the West had seen the Waffen SS established as a competent military force. Given that apart from

ABOVE LEFT: Total exhaustion during the battle for France.

ABOVE: *Totenkopf* soldiers celebrate the victory in France with looted champagne.

ABOVE RIGHT: The shield and cuff title of the Freiwilligen Legion *Flandern*.

RIGHT: Early recruits for the Freikorps *Danemark* march off for induction.

FAR RIGHT: Recruits for the Freiwilligen Legion *Norwegen* take the oath of loyalty.

Polizei, all the Waffen SS units were motorized, the German Army had deployed them in offensive, mobile operations, and as such they had seen some heavy fighting. The Waffen SS gained valuable experience from the campaign of 1940. SS soldiers were determined and tough, and generally carried out their orders with a sense of recklessness and this was reflected in their heavy casualties. In 18 days of contact with the enemy, including seven days of heavy fighting, *Totenkopf* suffered 1152 casualties, or slightly more than 10 percent of its combat strength. Bravery and recklessness did not compensate for professional skills. Although army commanders were critical of the Waffen SS, they were content to have them under command. But the ethos and ideology of the Waffen SS, combined with their frustration and blood lust, had resulted in them committing atrocities. The murder of unarmed British prisoners at Wormhoudt and Le Paradis might not have been ordered by Waffen SS officers but it was condoned.

Hitler was delighted with the performance of the Waffen SS, and in his speech to the Reichstag on 19 July included them in his fulsome praise of the Wehrmacht, 'the German Panzer Corps has inscribed for itself a place in the history of the world; the men of the Waffen SS have a share in this honor.' Himmler made sure the Waffen SS had its share of promotions and medals, including the award of six Knight's Crosses.

During June and July the Wehrmacht prepared for the invasion of Britain. In July Hitler declared his secret intention to fight a war against the Soviet Union, but the Wehrmacht continued to go through the motions of preparing to invade Britain. As part of the Wehrmacht expansion for the war in the East Hitler allowed the Waffen SS one new division and an increase in the strength of the *Leibstandarte*. Hitler was against a larger expansion of the Waffen SS because he still thought in terms of it being a *Staatstruppen-Polizei*, a militarized state police, and he wished it to remain a racial and ideological elite recruited from German volunteers. The army refused to allow the Waffen SS a larger share of recruits and Berger was forced to look outside the borders of the Reich. Two potential sources were open to Waffen SS recruiters, the *Volksdeutsche*, racial Germans living throughout Europe in considerable numbers, and Europeans of Germanic blood.

The attraction of both these groups was that they were not subject to Wehrmacht control. Berger's SS recruiters began to acquire *Volksdeutsche* volunteers in Romania and Hungary, and Germanic volunteers from the Western occupied countries were recruited into the Waffen SS. Norwegians and Danes were taken into the SS Regiment

Nordland while Dutch and Flemings went to the SS Regiment *Westland*. These men were following the trickle of Germanic volunteers who had joined the Waffen SS from 1938 after Himmler had approved their recruitment. Although initially there were insignificant numbers of either *Volksdeutsche* or Germanic volunteers, their recruitment set a precedent and at the same time eroded the prewar image of the SSVT.

In August 1940 Himmler created the SS *Führungshauptamt* (SS Main Operational Office) to act as the command headquarters of the Waffen SS and to negotiate on equal terms with the high commands of the three services and OKW. As chief of staff, and after 1943 as commander, Himmler appointed SS Brigadeführer Hans Jüttner. Jüttner, born in 1894, had been a bank clerk before serving as an officer on the Turkish front during World War I. Service in the *Freikorps* and membership of the Nazi Party and the SA prepared Jüttner before he joined the SSVT in 1935. With the SS *Führungshauptamt* Jüttner created a vast empire to administer the Waffen SS. One of the more specialist units attached to the SS *Führungshauptamt* was an SS Military Geologists Battalion.

From August 1940 until the spring of 1941 the Waffen SS was expanded and reorganized. Using German volunteers, *Volksdeutsche*, Germanics and manpower from the SS *Totenkopf* Regiments, Himmler was able to double the size of the Waffen SS field army. SSVT was renamed *Das Reich* and, under the command of Felix Steiner, a new division, *Wiking*, was created from drafts from other Waffen SS units and the SS Regiments *Nordland* and *Westland*. Throughout the war *Wiking* was to have a considerable number of Germanics serving in its ranks. Finally, Himmler formed Kampfgruppe *Nord* and an additional SS Infantry Regiment from available *Totenkopfstandarten* personnel.

Himmler still retained armed units in addition to the Waffen SS formations directly under army command: the *Ersatz* battalions of the Waffen SS and the remaining elements of the SS *Totenkopfstandarten*. The SS *Totenkopf* Regiments had originally been used for police duties in the occupied territories, but Himmler found they could be replaced by battalions of overage Order Police. The remaining SS *Totenkopf* Regiments were given infantry weapons and some transport, and formed into two SS infantry and one SS cavalry brigade. These SS brigades were to form part of Himmler's police army to be used for special duties behind the German lines during the

ABOVE LEFT: Vidkun Quisling (second from left) and SS Gruppenführer Hans Jüttner (fourth from left) inspect the Freiwilligen Legion *Norwegen*.

LEFT: *SS Kampfgruppe Nord* in Finland.

ABOVE: SS Gruppenführer Sepp Dietrich and SS Hauptsturmführer Max Wünsche with officers of a German mountain division in Greece, 1941.

war against the Soviet Union. As this war was racial and ideological Himmler was tasked with creating special forces to exterminate all Jews, communists and partisans. Himmler had three groups of forces at his disposal. The three SS brigades would operate in an independent role and eliminate Red Army stragglers and partisans, while *Einsatzgruppen*, drawn from the Security Police and SD, the Order Police and the Waffen SS, would undertake the extermination of Jews and communists in the army rear areas. Finally, regiments of Order Police would carry out the tasks originally conceived for the SS *Totenkopf* Regiments. During the course of the war Himmler would establish 30 SS Police Regiments and dozens of battalions of auxiliary police composed of Baltic peoples and Ukrainians.

In April 1941 Himmler formally listed some 163 separate units and organizations that he considered part of the Waffen SS. Included were the personnel of the staff and guard detachments of the concentration camps, and later the extermination camps, who were to wear Waffen SS uniforms and carry Waffen SS paybooks. By the end of the war some 35,000 of these guards were wearing the uniform of the Waffen SS, including Rudolf Hoess, the commandant of Auschwitz who had served under Eicke at Dachau before the war.

The veteran Waffen SS formations, *Leib-*

standarte, *Das Reich*, *Totenkopf* and *Polizei*, remained in France refitting and preparing for the invasion of Britain, and when that was postponed, for the invasion of the Soviet Union. Large-scale training exercises included mobile warfare, assault river crossings and fighting in villages and wooded areas. The Waffen SS like the army were told that the forthcoming campaign in the Soviet Union was racial and ideological, that the normal rules of war would not apply, and that the struggle would be brutal and hard. German soldiers were reminded that they were members of a racial elite and that the Soviet enemy was subhuman. In the *Totenkopf*, Eicke took political indoctrination very seriously and it carried equal weight with military training. Waffen SS commanders were determined that their men would be spiritually and physically prepared for the new campaign. Like their colleagues in the army, they were convinced that Operation *Barbarossa* would be a short, Blitzkrieg campaign similar to that in the West. However, events in the Balkans were to precipitate the move of certain army and Waffen SS units.

The Italian invasion of Greece in October 1940 had been a disastrous failure. The Greeks had put up a stout resistance and brought the Italian Army to its knees. Hitler was unhappy about this vulnerable southern flank which could endanger *Barbarossa*,

and in the first three months of 1941 16 German divisions were deployed to the Balkans, with the *Leibstandarte* moving in February. After British troops landed in Greece in March 1941, Hitler decided to settle matters. Yugoslavia was persuaded to align itself with the Axis powers, leaving the Greeks isolated apart from British support. To Hitler's fury the pro-Axis government in Yugoslavia was overthrown and replaced by a pro-British government. Plans to invade Greece were hastily amended to include Yugoslavia, and on 28 March *Das Reich* was moved from France to Romania to take part in the attack on Yugoslavia.

Das Reich formed part of the XLI Panzer Corps which crossed the Yugoslav border on 6 April. The Yugoslav armed forces were totally unprepared for modern warfare. Following a devastating Luftwaffe air raid, an assault detachment from *Das Reich* was the first German unit to enter Belgrade and accept the surrender of the capital. The *Leibstandarte* was to have a more challenging and satisfying campaign. Forming part of the XL Corps, *Leibstandarte* followed the 9th Panzer Division into Yugoslavia and within three days had captured Skoplje and

LEFT: *Das Reich* parades through Belgrade.

ABOVE: Kurt 'Panzer' Meyer in Greece.

LEFT: Local civilians greet Waffen SS soldiers during the Balkan campaign.

BELOW: Motorized troops of the Waffen SS in Russia, June 1941. The armored car is an SdKfz 222.

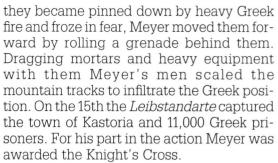

reached the Yugoslav-Greek border. The *Leibstandarte* was ordered to open the Klidi Pass, the gateway into Greece, which was defended by Australian troops. A battle-group under Sturmbannführer Fritz Witt began to attack the pass on 10 April. It took Witt's men three days of heavy fighting to seize the heights dominating the pass at a cost of 37 killed, 98 wounded and two missing. But the Klidi Pass had been opened and General Stumme, the corps commander, thanked the *Leibstandarte* for its efforts 'which resulted from the same unshakable spirit which the *Leibstandarte* constantly displays.'

The Germans pursued the Allies toward the Klissura Pass which was held by Greek troops. The Greeks placed demolition charges along the narrow road, making it impossible to bring up vehicles or artillery. Sturmbannführer Kurt 'Panzer' Meyer, commanding the *Leibstandarte*'s reconnaissance battalion, led his men forward. When they became pinned down by heavy Greek fire and froze in fear, Meyer moved them forward by rolling a grenade behind them. Dragging mortars and heavy equipment with them Meyer's men scaled the mountain tracks to infiltrate the Greek position. On the 15th the *Leibstandarte* captured the town of Kastoria and 11,000 Greek prisoners. For his part in the action Meyer was awarded the Knight's Cross.

The *Leibstandarte* was then ordered to send a force from Kastoria toward Elasson, southwest of Mount Olympus. By 20 April elements of the *Leibstandarte* had secured Katarra Pass and were amazed to receive a military delegation offering to negotiate the surrender of the Greek Armies of the Center and Epirus. Dietrich made his way forward and concluded an honorable capitulation with General Tsolakoglou. It was a heady moment for Dietrich, the former NCO, to accept the surrender of 16 Greek divisions. In fact Dietrich got into trouble with Hitler for accepting the Greek surrender without involving the Italians. For the Waffen SS, the campaign in the Balkans was over, but it had been a useful experience and enabled the *Leibstandarte* to receive a considerable amount of publicity.

By June 1941 the Germans had deployed 145 divisions in the East. A vast invasion force of seven armies, four panzer groups and three Luftwaffe air fleets stood poised in a line from the Baltic to the Black Sea. Numerically, the Waffen SS contribution was small, and its divisions were allocated to different army groups. In Finland, SS Kampfgruppe *Nord* and an SS Infantry Regiment were with General von Falkenhorst's Norwegian Army Command. The *Totenkopf* and *Polizei* were allocated to Field Marshal von Leeb's Army Group North; *Das Reich* to Field Marshal von Bock's Army Group Center and *Leibstandarte* and *Wiking* to Field Marshal von Rundstedt's Army Group South. Over 100,000 Waffen SS soldiers were serving with the field army. Behind them were Himmler's independent SS brigades, the *Einsatzgruppen* and battalions of Order Police which were ready to fight the racial war.

Despite this massive concentration of force against the Soviet Union there were serious flaws in German planning. Neither Hitler nor his generals could decide on how the Soviets were to be defeated. Instead of concentrating either on one major thrust against Moscow after destroying the bulk of the Red Army or, alternatively, seizing the Ukraine before redeploying northward, initial planning did not develop in detail beyond taking operations to Smolensk. The army's military planning for *Barbarossa* was totally inadequate. There was insufficient detailed information available concerning the political, military and economic strength of the Soviet Union and a failure to prepare for the problems of operating over a vast area quite different to that of Western Europe. Combined with an arrogant contempt for the Russians it was not a basis for a quick, decisive campaign.

In the early hours of 22 June 1941 the Germans launched *Barbarossa*. For the Waffen SS it was to be the beginning of a struggle which would enhance its military reputation but forever tarnish its image. Within a few weeks of the start of the campaign the Waffen SS found it was fighting a tenacious and formidable enemy across difficult terrain. In the north, *Totenkopf* and *Polizei* advanced through the Baltic states toward Leningrad. The *Totenkopf* was unpleasantly surprised by the fighting skills of the Red Army. Tenacious in defense, prepared to accept heavy casualties in counterattacks, they continued to fight even when cut off behind the German advance. On 6 July the *Totenkopf* began to fight its way through the Stalin Line and Eicke was wounded when his car hit a mine. After 16 days of fighting the division had lost 82 officers and 1626 NCOS and soldiers killed, nearly 10 percent of its combat strength. General Manstein criticized the *Totenkopf* for its heavy casualties which were disproportionate to its relatively modest gains.

In July the *Totenkopf* advanced through the swamps and gloomy forests southwest of Lake Ilmen on the approach to Leningrad. The fighting was exhausting for the SS, advancing during the day and repulsing Soviet attacks at night. The Soviets infiltrated SS positions, snipers killed officers and despatch riders, and groups of partisans and Red Army stragglers ambushed supply units. Far from the Red Army being close to the point of collapse, the *Totenkopf* discovered that its ability to launch counterattacks forced them on to the defensive. Manstein used the *Totenkopf* as one of his spearhead divisions to cross the Pola River at the end of August. When Eicke returned to command his division on 21 September he was shocked by the physical appearance of his men. Over the following week the *Totenkopf* faced a major Soviet attack along the Pola designed to rip into the flank of the German forces besieging Leningrad. The *Totenkopf* came under intense artillery fire and air attacks, and its infantry were literally engulfed by human tidal waves. By the evening of the 26th one battalion had lost every officer killed, including four succes-

LEFT: Blitzkrieg in Russia – German tanks resume the advance.

RIGHT: *Polizei* soldiers equipped with an MG34 machine gun prepare to repel a Soviet counter attack.

LEFT: A 7.92mm machine-gun team of the *Wiking* Division in action in Russia.

ABOVE: Waffen SS engineers repairing a bridge protected by a 2cm antiaircraft gun.

sive commanders. During the next day the Soviets attacked with 100 tanks and three infantry divisions but failed to break the *Totenkopf.*

The primary reason for the success of the *Totenkopf* during this battle was the fighting spirit of the individual SS soldier. Hard, ruthless, believing himself to be racially superior to his enemy, commanded with fanatical determination by Eicke, he gained the unqualified respect of Generals Manstein and Busch. Finding their antitank guns to be generally ineffective against the Soviet T-34 tanks, Eicke created special 'tank annihilation squads' of men armed with mines, grenades and petrol bombs. Hauptsturmführer Max Seela of the *Totenkopf* engineer battalion was awarded the

Knight's Cross for leading his men in destroying seven Soviet tanks. His technique was to leap on the tank while it was moving, placing a charge against the turret which he then detonated with a grenade.

An outstanding example of bravery and obedience to orders was given by SS Sturmann Fritz Christen, an antitank gunner. On 24 September every SS soldier was killed in the battery except Christen. For three days, completely cut off from any help, he remained in the battery serving the guns on his own. Finally, when relieved by *Totenkopf* troops, Christen was found to have personally destroyed 13 Soviet tanks and killed nearly 100 soldiers. Subsequently, Christen was personally awarded the Knight's Cross by Hitler.

The SS man's readiness to die rather than retreat or appear weaker than his racial enemy, his lust for fighting, his willingness to hold out against impossible odds were the fighting characteristics of the better Waffen SS units throughout the war. In October and November 1941 the weather worsened and Eicke's men suffered from the extreme cold. Casualties since June had been heavy with 8993 dead, wounded or missing, or nearly 50 percent of the division. Despite these losses Army Group North regarded the *Totenkopf* as one of its best formations, one crucial to the security of the right flank south of Lake Ilmen.

In Army Group South Sepp Dietrich's *Leibstandarte* had been attached to General von Mackensen's III Corps for the offensive in the Ukraine. Like the *Totenkopf*, the *Leibstandarte* soon realized how formidable the Red Army was either in defense or attack. Soviet tactics were primitive as one *Leibstandarte* soldier observed: 'The counter-

attack . . . was launched while we were still getting our breath back. Their infantry came in mounted on open lorries which swayed from side to side with the speed. It looked as if all the Ivans on the trucks were standing up and firing their guns at us. The whole thing was quite primitive. The lorries just drove straight at our positions ... a shell hit a lorry and killed many of the infantry mounted on it, but the others sprang over the side and charged us on foot ... There was no cover at all . . . They had no hope of reaching our positions, but they still came on ...' During the first three weeks of July the *Leibstandarte* suffered 383 casualties and had 100 vehicles destroyed. It was mentioned in a special army corps order of the day for its part in capturing 100,000 Soviet prisoners in the the Uman Pocket.

As a motorized unit, the *Leibstandarte* was able to keep up with the advancing panzers. In the first week of September the *Leibstandarte* crossed the Dnieper and advanced across the Nogai steppe. Kurt Meyer's reconnaissance battalion led the way toward the Perekop Isthmus, noting that 'this is true desert country. Movement is visible for miles; clouds of choking, red-brown dust hang over our moving columns and pinpoint our exact positions. Paradoxically, the only sign of life are the dead tree trunks of telegraph poles. Without them it would be difficult to orientate oneself.' Throughout these months Hitler took a personal interest in the movements of the

Leibstandarte and made sure it was used in the most effective and glamorous manner.

In October *Leibstandarte* took part in the successful encirclement of 100,000 Red Army troops along the Sea of Azov and was moved forward to participate in the capture of Rostov. But stiffening Soviet resistance and the onset of rain slowed the advance. The temperature dropped to −20°C and the *Leibstandarte* faced determined Soviet counterattacks. One *Leibstandarte* soldier wrote: 'It is not possible in words to describe winter on this front. There is no main battle line, no outposts, no reserves.

TOP: Ukrainian peasants welcome Waffen SS soldiers.

ABOVE: SS troops display a captured Red Army banner. Its legend reads, 'Fight for Lenin and Stalin — be prepared!'

TOP: 'General Winter' comes to the aid of the Red Army. Here, a well-clothed machine-gun detachment goes into action.

ABOVE: Fighting around Lake Ilmen to the south of Leningrad. Note the rarely-seen camouflaged face masks.

Just small groups of us depending upon each other to hold defended points . . . We are in the sunny south, how frightful it must be for the comrades up North. Here life is paralyzed . . . you would never believe the lavatory procedures . . . And the food . . . we live on a sort of thick soup made of ground buckwheat and millet. We have to strip the fallen, theirs and ours, for warm clothing. I don't think I will ever be warm again and our tame Ivans say that this is a mild winter. God preserve us.'

Without winter clothing, short of heavy weapons and ammunition, the *Leibstan-* darte held on with grim determination. The Soviets attacked 'in masses so great as to numb the senses. They had to pick their way through the dead of the other assaults who are still unburied. We drove them off — how easy it seems to write this . . . and when they had gone back across the ice the whole area to both flanks and in front of our positions was carpeted with dead. They were dead all right . . . the wounded die quickly; the blood freezes as it leaves the body and a sort of shock sets in which kills. Light wounds that heal in three days in summer kill you in the winter.' At the end of November the *Leibstandarte* reported 31 deaths from frostbite.

Faced with these Soviet counterattacks Rundstedt ordered a withdrawal from Rostov to consolidate a line behind the Mius River. A furious Hitler dismissed him, but Dietrich personally told Hitler that the withdrawal from Rostov had not been due to any deficiencies in the leadership of Army Group South. Dietrich told Hitler that the *Leibstandarte* had lost 50 percent of its manpower and had only 15 percent of its vehicles operational.

In December 1941 General von Mackensen wrote to Himmler extolling the fighting qualities of the *Leibstandarte*, 'Every division wishes it had the *Leibstandarte* as its neighbor, as much during the attack as in defense. Its inner discipline, its cool daredeviltry, its cheerful enterprise, its unshakable firmness in a crisis (even when things

become difficult or serious), its exemplary toughness, its camaraderie (which deserves special praise) – all these are outstanding and cannot be surpassed.' Sepp Dietrich was singled out for praise by Nazi propaganda. *Das Schwarze Korps*, the SS journal, described Dietrich as 'the father of his men, as the model for his unit commanders, a hard soldier with a strange tender heart for his comrades.' Hitler compared Dietrich with German heroes such as Frundsberg, Zeiten and Seydlitz, describing him as 'simultaneously cunning, energetic and brutal,' and significantly, 'one of my oldest companions in the struggle.' General Wohler, commanding the Eighth Army, said of the Waffen SS troops under his command that they had 'stood like a rock in the army,' and withstood Soviet attacks 'with unshakable fortitude.' The fighting qualities of

the Waffen SS even won the respect of a Soviet general captured by Army Group Center, who said *Wiking* had shown greater fortitude than any other formation on either side and that the Soviets had been relieved when it was replaced by another division.

But there were tensions between the army and the Waffen SS, with army generals still critical of the excessively heavy casualties of the Waffen SS, and Waffen SS commanders like Eicke accusing the army of 'burning up' their units in impossible situations. And the military record of the Waffen SS was not one of outstanding bravery and operational effectiveness on every occasion. To Himmler's annoyance the SS Kampfgruppe *Nord* suffered a defeat in July 1941 while attacking a Soviet position at Salla on the Finnish Front. The Soviets counterattacked and many SS soldiers ran

ABOVE: Waffen SS cavalry take a break from operations.

ABOVE RIGHT: Waffen SS and army soldiers taking 'snapshots' of executed Soviet partisans.

RIGHT: A Red Army tank crew surrender to an SS NCO.

away in panic or surrendered, the situation only being saved by the action of the Finns and an army division. General von Falken-horst having lost faith in the fighting qualities of the kampfgruppe divided its battalions among his Finnish and German Army formations. Poor leadership and the fact that the SS men were inexperienced former concentration camp guards and overage SS reservists were the causes of this Waffen SS humiliation.

The war in the East was characterized by both sides violating the rules of war. The Germans had planned a racial and ideological war against both soldiers and civilians alike, and the Soviets fought a ruthless war at the front and in the German rear areas. Like the army, the Waffen SS shot many prisoners out of hand, and responded in kind to Soviet atrocities. A fortnight after the beginning of the campaign the *Wiking* Division shot 600 Jews in Galicia for Soviet crimes. In July a company of engineers from *Das Reich* were temporarily cut off behind the Soviet lines. When the survivors were finally rescued, it was found that the Soviets had shot all those wounded or taken prisoner.

Behind the German front line the *Einsatz-gruppen*, the SS cavalry and infantry brigades and the Order Police shot tens of thousands of Red Army prisoners, Jews and partisans. In *Einsatzgruppe* A, which followed Army Group North through the Baltic

States toward Leningrad, 34 percent of the personnel were Waffen SS. This one *Einsatzgruppe* of 990 men had 'liquidated' 249,420 Jews by the winter of 1941-42. The distinction between front and rear became blurred, with Waffen SS and army personnel in transit or resting behind the front line giving assistance to the *Einsatzgruppen* and Order Police in killing civilians and combating partisans, and when there was a crisis at the front Himmler's police army deployed for combat duties.

For both the army and the Waffen SS the winter of 1941-42 proved to be a test to destruction. At the beginning of December 1941 the Germans had exhausted their strength outside Leningrad, Moscow and Rostov. On the 6th 12 Soviet armies under Marshal Zhukov attacked Army Group Center on a 500-mile front. At first there was panic and paralysis until Hitler took personal command, dismissing generals and issuing a 'no retreat' order. By mid-January a combination of Hitler's order and the fighting qualities of the German soldier had halted Zhukov's offensive. *Das Reich* proved to be invaluable as a bulwark against Soviet attacks. For one month, in a temperature of −52°C, the soldiers of SS Regiment *Der Führer* held a vital sector at Rzhev on the Volga. General Model, commanding the Ninth Army, finally relieved the regiment to find that only 35 out of 2000 men were still fit for duty.

Within 24 hours of the counteroffensive against Army Group Center, the Soviets attacked Army Group North, achieving a breakthrough between Lake Ilmen and Seliger, and surrounding the *Totenkopf* and several army units in the Demyansk Pocket. Between January and October 1942 the *Totenkopf* was the nucleus of the German defensive position in the Demyansk Pocket. The original *Totenkopf* Division was all but destroyed in this fighting. Hitler and his generals recognized its critical importance in holding together other units, and Eicke was forced to see his division bleed to death. Conditions were so bad that at one point there were three known *Totenkopf* desertions to the Soviets, a phenomenon extremely rare amongst Waffen SS units. Finally, the remnants of the *Totenkopf* were withdrawn in October 1942 and joined other veteran Waffen SS units which were in the pricess of being refitted as a panzergrenadier division.

BELOW: Waiting for the order to advance.

ABOVE Primitive roads in Russia caused problems, especially with the onset of wet or wintery weather.

RIGHT: The butcher's bill – dead Waffen SS at Lake Ilmen.

IN 1942 Hitler was still optimistic of an early German victory in the war, but although he was very impressed by the fighting spirit shown by the Waffen SS on the Eastern Front, he was not prepared to authorize a major expansion. Hitler still believed that the Waffen SS had to remain relatively small and racially cohesive, but he did permit the reorganization and reinforcement of existing formations. Hitler had made his views clear in January 1942 when he said: 'The SS shouldn't extend its recruiting too much. What matters is to keep a very high level. This body must create upon men of the elite the effect of a lover. People must know that troops like the SS have to pay the butcher's bill more heavily than anyone else – so as to keep away the young fellows who only want to show off.' Hitler retained in his mind the primary role of the Waffen SS, 'In peace time

now backed Himmler in his struggle for manpower with the other services, and the Waffen SS was permitted to recruit three times its normal quota from the class of 1924. But insufficient volunteers came forward and for the first time the Waffen SS was forced to conscript German youth. The mobility and firepower of the new panzergrenadier divisions was increased with the acquisition of tanks, assault guns, and additional antitank and antiaircraft battalions.

Even with Hitler's support, the Waffen SS faced serious problems in recruiting sufficient numbers of young men who reached its physical and racial standards. Himmler's head of recruiting, SS Gruppenführer Berger, had continued to exploit two manpower sources outside the frontiers of Germany, the Germanics and the *Volksdeutsche*. In addition to the Germanic

CHAPTER 3

The Führer's Firemen 1942-45

it's an elite police, capable of crushing any adversary. It was necessary that the SS should make war, otherwise its prestige would have been lowered.'

Himmler had already begun the task of reorganizing the Waffen SS. Following the humiliating defeat of Kampfgruppe *Nord* in July 1941, he drafted experienced officers and NCOs from other units and in August it was expanded into a division. In the summer of 1942 the SS cavalry brigade was turned into the SS Kavalleriedivision *Florian Geyer*. During 1942, *Leibstandarte, Das Reich, Totenkopf* and *Wiking* were all expanded into panzergrenadier divisions, and Hitler agreed to the establishment of an SS corps commanded by Hausser. Failure to achieve a decisive victory in Russia in the summer of 1942 and defeat in North Africa persuaded Hitler to allow further expansion. In December Hitler authorized two new panzergrenadier divisions, *Hohenstaufen* and *Frundsberg*, and a second corps. Hitler

volunteers who were coming forward and being sent to *Wiking*, in June 1941 Hitler gave Himmler permission to raise national legions from each German-occupied nation in Western Europe who could participate in the "Battle against Bolshevism." Berger established four legions, recruited from Danes, Norwegians, Dutch and Flemings: the *Freikorps Danemark*, and the *Freiwilligen Legionen Norwegen, Nederlande* and *Flandern*.

The initial results were disappointing and none of the legions ever achieved regimental strength. Poor leadership and abysmal treatment at the hands of the German instructors failed to inspire potential recruits, who were quick to learn that although they wore Waffen SS uniforms and served with Waffen SS units they were not Waffen SS soldiers. By November 1941 the Dutch and Flemish legions were partially trained and were attached to the 2nd SS Infantry Brigade on the Leningrad front

LEFT: Members of the SS Kavallerie Division *Florian Geyer* enjoy a cigarette break.

FOR DANMARK!
MOD BOLCHEVISMEN!

MED WAFFEN-SS OG
DEN NORSKE LEGION
MOT DEN FELLES FIENDE......

MOT BOLSJEVISMEN

THESE PAGES: Some indication of the attempts to raise Waffen SS 'legions' in Western Europe. Short of recruits and attempting to create a 'Pan-European' force, the SS launched a propaganda 'blitz' to attract the dissatisfied young of several countries. Many rallied to the call – out of conviction or a sense of adventure.

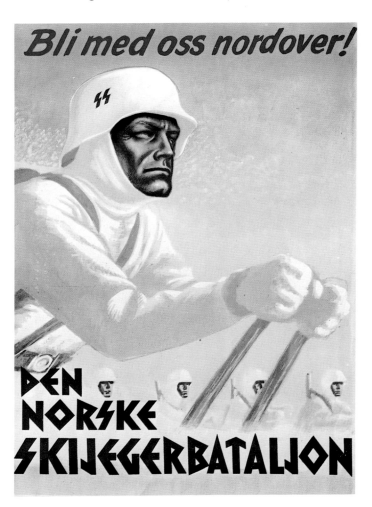

Bli med oss nordover!

DEN NORSKE SKIJEGERBATALJON

Til vakt ved Nordens grense mot øst!

SS-SKIJEGERBATALJON
NORGE

where they were badly mauled. The *Freikorps Danemark* was sent as a replacement unit to the *Totenkopf* in May 1942 and then transferred and joined the 1st SS Infantry Brigade along with the Norwegians. By the summer of 1942 there were perhaps some 4000 Germanic SS legionnaires serving with the Waffen SS in Russia.

There was considerable friction between the volunteers and their instructors and Himmler was forced to issue a series of directives to deal with the worst cases. In late 1942 Himmler decided to concentrate the legions in a newly created SS panzergrenadier division *Nordland*. Ironically, in the last two years of the war recruitment from Western Europe improved as Himmler accepted non-Germanics and as many collaborators sought shelter in the Waffen SS from the Resistance and Allied retribution. By 1945 there were nominal divisions of Dutchmen in *Nederlande,* Flemings in *Langemarck*, Walloons in *Wallonien*, Frenchmen in *Charlemagne* and Italians in a Waffen-Grenadierdivision der SS.

As many as 125,000 West Europeans served in the Waffen SS, and nearly half of them enlisted before 1943-44. The largest national contingent was provided by the Dutch at 50,000, then 20,000 Walloons, 20,000 Flemings, 20,000 French and 6000 each of Danes and Norwegians. These volunteers joined the Waffen SS for a variety of reasons. The largest group joined out of a sense of adventure and boredom or because they were unemployed. This group was closely followed by those who belonged to ideological and national organizations that wished Germany to win the war and believed that by volunteering they would ensure their party and national group would have a privileged place in Hitler's New Order. Some joined because they were fiercely anticommunist. In the final 18 months of the war a large group joined to avoid compulsory labor service or criminal prosecution.

The average volunteer was working class, apolitical and immature. Typical was the 20-year-old Dutchman Gerardus Mooyman, who volunteered much to the shame of his devoutly Catholic family. Serving with the Dutch legion he was to win the Knight's Cross south of Lake Ilmen in February 1943. The best of them served in units which were led by their own officers like Leon Degrelle and his Walloons, or commanded by Waffen SS generals like Felix Steiner who promoted the concept of a community of European nations led by Germany in the fight against Bolshevism. But Steiner's idealistic and näive concepts found little favor either with Hitler or Himmler who were quite clear that a future Europe would be dominated by Germany. Himmler was prepared for pragmatic reasons to tolerate within the Waffen SS those who encouraged pan-European ideas, but never once did he deviate from his Germanic ideology.

The *Volksdeutsche* provided the other

ABOVE: Felix Steiner (left) addresses the Finnish Battalion.

BELOW: Leon Degrelle of the Panzerbrigade *Wallonien.*

BOTTOM: Arthur Quist, commander of the Freiwilligen Legion *Norwegen.*

ABOVE: SS Obergruppenführer Artur Phleps, commander of the *Prinz Eugen.*

BELOW: SS Sturmann Gerardus Mooymann, Knight's Cross holder of the Freiwilligen Legion *Niederlande.*

large potential reservoir of manpower for the Waffen SS. After the defeat of Yugoslavia Berger promised Himmler that he would be able to recruit enough *Volksdeutsche* to man a new Waffen SS division. In March 1942 *Prinz Eugen* was established, but to Berger's embarrassment insufficient numbers of *Volksdeutsche* volunteered and so he was forced to resort to conscription. Command of the *Prinz Eugen* was given to Artur Phleps, a Romanian *Volksdeutsche* who had served as a general in the Romanian Army. Organized as a mountain division and armed with captured French, Czech and Yugoslav weapons, *Prinz Eugen* served throughout the war in Yugoslavia conducting antipartisan operations with great brutality.

After 1942 Berger was able to put immense pressure on Germany's allies – Croatia, Slovakia, Romania and Hungary – to allow him to recruit from their *Volksdeutsche* minorities. By the summer of 1944 more than 150,000 *Volksdeutsche* were serving with the Waffen SS. In terms of numbers Berger had succeeded in providing Himmler with racially acceptable recruits. But Jüttner at the SS *Führungshauptamt*, instructors in the *Ersatz* battalions, and divisional commanders like Dietrich and Eicke were far from satisfied. They found many of the *Volksdeutsche* totally unsuitable – physically unfit, mentally subnormal, unable to speak German, and uncommitted to Germany and National Socialism. By the end of the war hardly a Waffen SS unit was without a sizeable contingent of *Volksdeutsche*. In April 1945 British soldiers were surprised to find that many of the guards wearing Waffen SS uniform at Bergen-Belsen concentration camp were Hungarian *Volksdeutsche*. In assessing the military proficiency of Waffen SS divisions it is possible to establish a correlation with the percentage of *Volksdeutsche* in their ranks – the higher the percentage, the lower the combat performance.

Total war, manpower shortages and institutional rivalry with the army all helped to persuade Himmler to relax his standards even further, so that by 1943 even racial subhumans of only yesterday were eligible to become "honorary aryans" – at least for the duration of hostilities. In 1942 Berger persuaded Himmler to establish Baltic SS Legions recruited from Estonians and Latvians who would be prepared to fight against the Soviet Union to secure their national independence. Lithuanians were excluded by Himmler because they were considered racially inferior and unreliable. In 1943 conscription was introduced in Estonia and Latvia and three *Waffen-Grenadierdivisionen der SS* were formed.

The guerrilla war in Yugoslavia threatened to draw in German divisions from other theaters, and Himmler was able to justify to himself the recruitment of Bosnian Moslems. With their traditional hatred of the Christian Serbs and the atheistic communists they appeared suitable recruits for the war against Tito's partisans. After initial difficulties with the Croat regime, Berger's recruiters had attracted enough Moslem recruits to form the basis of what after March 1943 would become the *Waffen-Gebirgsdivision der SS Handschar.*

Recruiting Moslems into a Waffen SS formation completely contradicted Himmler's racial and ideological standards. Organized along the lines of Moslem units which had served in the Imperial Austro-Hungarian Army, *Handschar* had German, Austrian and *Volksdeutsche* officers and NCOs. Moslem recruits were given special privileges relating to diet and religious observances. With modified Waffen SS uniforms – all ranks in service dress wore a field-gray fez

with a tassle and suitable SS insignia – *Handschar* was one of the more colorful and bizarre Waffen SS divisions, and in the final analysis a liability and failure. It mutinied during training in France and further acts of indiscipline in Yugoslavia persuaded Himmler to disband the unit in October 1944. Despite the unfortunate example set by *Handschar*, Himmler raised two further Balkan divisions: *Waffen-Gebirgsdivision der SS Skanderbeg* recruited from Albanians and *Waffen-Grenadierdivision der SS Kama* recruited from Croatians. Neither unit was to distinguish itself in combat.

From the beginning of *Barbarossa* the German Army and the higher SS and police leaders had recruited former Red Army personnel and local Russians, Ukrainians, Estonians, Latvians and Lithuanians into security and auxiliary police battalions. Himmler was prepared to use these *wilde Völker* ('savage peoples') to kill Jews and fight partisans. By 1943 some of Himmler's last scruples concerning recruitment to the Waffen SS were abandoned when he permitted the raising of the SS *Freiwilligen-Division Galizien* from Ukrainian nationalists. After 1944 two further *Waffen-Grenadierdivisionen der SS* were formed from the flotsam of Russian collaborators, and eventually an SS Cossack Corps. On the fringes of military absurdity was an SS Indian Legion recruited from Indian Army prisoners and a British Free Corps consisting of two dozen or so former prisoners and British fascists.

Apart from the three divisions recruited from Estonians and Latvians, the rest of those from Eastern Europe were useless. Even in antipartisan operations such units were a liability due to their indiscipline and propensity for looting. Many of these 'divisions' were so in name only, and apart from a cadre of German officers and NCOs, lacked any military experience. Himmler wisely kept their existence secret from Hitler, who, when he discovered them on the establishment in 1945, was both furious and contemptuous. He ridiculed their military effectiveness and railed against the absurdity of giving these units weapons and equipment when German units had none, and questioned their loyalty to Nazi Germany. It was one of the ironies for Hitler that some of the last defenders of central Berlin in April 1945 were Danes and Norwegians from *Nordland*, Frenchmen from *Charlemagne* and Latvians from the 15th *Waffen-Grenadierdivision der SS*.

While *Leibstandarte, Das Reich, Totenkopf* and *Wiking* were being reconstituted as panzergrenadier divisions, the German

Army suffered a catastrophic defeat in Russia. The summer offensive of 1942 had forced the Red Army back to the Volga, but at Stalingrad the Germans had become overextended and were surprised by a Soviet counteroffensive which resulted in the encirclement of the Sixth Army. Refused permission to break out on Hitler's orders, the Sixth Army fought on waiting for relief that never came, and was forced to surrender at the end of January 1943. The whole of the German position in southern Russia appeared to be in danger as the Red Army advanced to the Dnieper.

In January 1943 Hausser's SS Panzer Corps was ordered to the Eastern Front. *Leibstandarte* and *Das Reich* were sent immediately and subordinated to Army Detachment Lanz which had the task of holding Kharkov. As remnants of German, Italian and Hungarian divisions fled westward, Hausser's SS Panzer Corps prepared to counterattack and halt the Soviet advance. Dietrich, commanding a composite Kampfgruppe, led his men into the attack in temperatures of −20°C. *Ad hoc* Kampfgruppen from the *Leibstandarte* and *Das Reich* desperately attempted to stop the Soviets.

Sturmbannführer Joachim Peiper, commanding a battalion of the 2nd Panzergrenadier Regiment of the *Leibstandarte*, led his men some 25 miles into Soviet-held territory to rescue the remnants of the encircled 320th Infantry Division which was encumbered with 1500 wounded. Peiper's battalion formed a protective screen and escorted the division back to the Donetz

ABOVE LEFT: Officers of the 23rd Waffen-Gebirgsdivision der SS *Kama.*

LEFT: *Handschar* artillery in training.

ABOVE: Cossack volunteers inthe German Army. Cossacks were also incorporated into the Waffen SS.

RIGHT: Bosnian Moslems of *Handschar* reading Nazi propaganda.

LEFT: Hungarian *Volksdeutsche* serving with the Waffen SS in Russia.

and safety. By 14 February the Soviets had penetrated into Kharkov and the SS Panzer Corps was in danger of being encircled. In spite of Hitler's categorical orders to the contrary, Hausser ordered a withdrawal from Kharkov thus saving his corps. Hitler was furious but Hausser retained his command. Dietrich, when asked what reserves he had, replied, 'Behind me are 400 kilometers of wind.'

Field Marshal von Manstein, commander of Army Group South intended using the SS Panzer Corps as part of a great pincer attack which would converge on and destroy the Red Army's advance to the Dnieper. With the arrival of *Totenkopf*, Hausser's SS Panzer Corps formed the northern point of the pincer with General Hoth's Fourth Panzer Army in the south. When the pincer closed the Soviet Sixth Army would be trapped and destroyed. On 19 February Manstein's counterattack began and within a week the Germans had destroyed the Soviet Sixth Army, including the capture or destruction of 615 tanks, 400 howitzers and 600 antitank guns. For the soldiers of the SS Panzer Corps it was like a field day with live ammunition and moving targets. Rumbling over the frozen steppe, the SS motorized columns often fired on the retreating Russians at distances of less than 30 yards. For the *Totenkopf* the victory was marred by the death of Eicke, shot down and killed in a light aircraft. Eicke became a Nazi hero and martyr, and Hitler renamed the *Totenkopf*'s panzer regiment *Theodor Eicke* in his honor.

Following the destruction of the Soviet Sixth Army Hausser deployed *Das Reich* and *Totenkopf* to lure forward the main body of the Red Army forces defending Kharkov, while sending *Leibstandarte* northwest of the city to a blocking position. On 10 March *Leibstandarte* moved into the city and in five days of fierce street and house-to-house fighting forced the Soviets out of Kharkov. During the German withdrawal and recapture of Kharkov, thousands of Soviet civilians were shot and this was blamed on the SS Panzer Corps. During the same period the SS Panzer Corps lost 11,500 casualties, with the *Leibstandarte*'s casualties alone standing at 4500.

Hitler now agreed to the formation of two more SS panzer corps. Hausser's corps was demoted to the II SS Panzer Corps, while a new I SS Panzer Corps was formed from the *Leibstandarte* and the newly established panzergrenadier division *Hitlerjugend*. The III SS (Germanic) Panzer Corps, commanded by Felix Steiner, was formed around *Wiking* and *Nordland*. *Leibstandarte* was to remain with Hausser until the conclusion of Operation *Zitadelle*, Hitler's summer offensive against the Soviet salient at Kursk. Sepp Dietrich was promoted to command the new I SS Panzer Corps.

In spite of Hitler's high opinion of him as a soldier, Dietrich's military skills barely enabled him to carry out the responsibilities of a divisional command, let alone that of a corps. Unlike Hausser, he had no professional training or command experience. SS Obergruppenführer Willi Bittrich recalled, 'I once spent an hour and a half trying to explain a situation to Sepp Dietrich with the aid of a map. It was useless. He understood nothing at all.' Hausser's caustic comment was, 'Ordinarily he would make a fair sergeant-major, a better sergeant and a first-class corporal.' Himmler and Jüttner noted Dietrich's shortcomings, and a qualified army general staff officer, Fritz Kraemer, was transferred to be his chief of staff.

Hitler issued a directive for *Zitadelle* in April 1943 which outlined the aim of the operation which was to destroy Soviet forces in central Russia. He planned using two powerful panzer armies on either side of the Soviet salient at Kursk, which would converge on each other along a north-south line. II SS Panzer Corps along with XLVIII Panzer Corps formed part of General Hoth's Fourth Panzer Army which would attack west of Belgorod at the southern end of the salient. Hausser's corps had received replacement manpower and new Tiger and Panther tanks which gave it an establishment of 343 tanks and 195 self-propelled guns. Disagreement between Hitler and his generals over operational objectives and Hitler's procrastination over the armored requirements for the offensive meant that *Zitadelle* was delayed until 5 July. The Soviets were well aware of the aim and scale of *Zitadelle* and had established a defensive position 95 miles in depth, consisting of bunkers, minefields and large mobile armored formations.

Following a Luftwaffe and artillery bombardment the ground forces of *Zitadelle* moved forward. A soldier in the *Leibstandarte* observed, 'our leading Tiger sections roar away and vanish almost completely in the peculiar silver-gray tall grass which is a feature of the area ... Our mine-lifting teams mark the position of Ivan's mines by lying down alongside them, then using their bodies to mark a gap in the field.' By the evening of the 7th II SS Panzer Corps had advanced some 30 miles. The numbers of Soviet tanks and guns destroyed, the thousands of dazed prisoners, the positions captured all gave the impression of victory. In reality, the SS had only just begun to encounter the main Soviet defensive position.

The progress made by the SS Panzer Corps had alarmed the Soviets, and on 12 July they launched a major armored counterattack. Against the 272 tanks and self-propelled guns of the SS were 850 Soviet tanks and self-propelled guns. The two formations clashed in a three-mile square area to the west of the Belgorod-Kursk railway line. For eight hours there was a major tank battle of such intensity that SS Tigers and Panthers and Soviet T-34s blazed away at each other at point-blank range, ramming each other when ammunition was exhausted. The SS Panzer

Corps absorbed tremendous punishment before the Soviets broke off their attack on the 14th. Although the Germans were to claim the destruction of 663 Soviet tanks and self-propelled guns, all the German divisions had lost over half their tanks and vehicles. Effectively *Zitadelle* had ground to a halt, and the crisis in the Mediterranean following the Allied invasion of Sicily gave Hitler an excuse to terminate operations.

Although Kursk had been a failure, Hitler had been impressed by the achievements of the Waffen SS. Over a period of six months, it had stabilized the front, recaptured Kharkov and made the deepest penetration of the Kursk salient before destroying the impetus of the Soviet counterattack. The contrast in Hitler's mind with the performance of the German Army could not have been greater. Unfairly ignoring the army's successes, he only thought of generals who disobeyed his orders and divisions which failed to achieve their objectives. Hitler was convinced that the success of the Waffen SS was due to their obedience to orders and the fact that they were a supposed racial and ideological elite.

After Kursk Hitler gave the Waffen SS priority in new weapons and equipment and authorized the establishment of more Waffen SS divisions. In October 1943 *Leibstandarte*, *Das Reich*, *Totenkopf*, *Wiking* and the newly established *Hohenstaufen*, *Frundsberg* and *Hitlerjugend* were designated SS panzer divisions. After 1943, with support from the SS Panzergrenadier Divisions *Polizei*, *Nordland*, *Reichsführer SS*, *Götz von Berlichingen* and *Horst Wessel*, Waffen SS panzer divisions were the 'Führer's Firemen,' moved from corps to corps and front to front to stabilize the situation and launching decisive counterattacks.

Although constituting less than five percent of the Wehrmacht in 1944, the fighting power of the Waffen SS was much greater. Seven out of 30 panzer divisions and six of the 17 panzergrenadier divisions were Waffen SS. This proportion was an indication of Hitler's confidence in the fighting power of the Waffen SS. The operational and tactical importance of these divisions was increased further by their numerical strength, the fact that they usually had 20 percent more armored vehicles than their army equivalent, and that these were often the latest weapons. Apart from the 12 core panzer and panzergrenadier divisions, the remaining Waffen SS divisions were of indifferent value, being understrength, cobbled together from a mismatch of personnel, led by police generals and badly equipped.

Although Hitler had intended sending II

BELOW: *Leibstandarte* panzers during the capture of Kharkov, 1943.

SS Panzer Corps to Italy, a Soviet offensive along the Mius River on 25 July forced him to retain *Das Reich* and *Totenkopf*. Only the *Leibstandarte* was sent to Italy, and for three months had a quiet time apart from some antipartisan work. Hitler used II SS Panzer Corps as his mobile fire brigade on the Eastern Front. At the end of July it saved the front on the lower Donetz, and then in August returned to the Belgorod area to counter a new Soviet offensive. For one week the corps acted as a breakwater against Soviet attempts to break through toward the Dnieper. Even when the Germans were forced to withdraw from Kharkov, *Das Reich* and *Totenkopf* launched counterattacks which slowed down the Soviet onslaught. Under immense Soviet pressure, II SS Panzer Corps and the army's *Grossdeutschland* Division covered the withdrawal of the Eighth Army across the Dnieper. Throughout the autumn of 1943, *Das Reich, Totenkopf* and, after November, *Leibstandarte*, formed the backbone of the German defenses in southern Russia. During the retreat of over 45 miles the Waffen SS divisions prevented a decisive Soviet breakthrough.

At the end of January 1944 the Soviets succeeded in encircling over six German divisions of Army Group South in the Cherkassy-Korsun Pocket. Among the 56,000 trapped Germans was *Wiking* and the SS brigade *Wallonien*. Manstein immediately prepared a relief force led by the *Leibstandarte*. Under appalling weather conditions, *Leibstandarte* and *Wiking* succeeded in establishing contact, and although nearly all their equipment was lost some 35,000 German troops escaped from the pocket. II SS Panzer Corps, now consisting of *Hohenstaufen* and *Frundsberg*, stopped a Soviet offensive in March along the Dniester. Army generals appreciated the crucial role played by the Waffen SS, and spoke of its divisions in glowing phrases such as 'stood like a rock in the Army, while the enemy broke through

ABOVE: *Leibstandarte* panzergrenadiers at Zhitomir, December 1943.

LEFT: Waffen SS panzergrenadiers ride into battle on a Panzer IV.

RIGHT: Sepp Dietrich (right) relaxing at Berchtesgaden, 1943.

BELOW: Mussolini being rescued by Otto Skorzeny from Gran Sasso, Italy, in September 1943.

in neighboring sectors' and 'a lightning sword of retribution with unshakable fortitude.' But all the Waffen SS had achieved was to buy time for Hitler in Russia. On every front Germany was on the defensive and the initiative lay with the Allies.

In the spring of 1944 Hitler withdrew the exhausted SS panzer divisions from Russia. Each division was reduced to a kampfgruppe of a few thousand men, and had to be rebuilt with new drafts of manpower and weapons. *Totenkopf* and *Wiking* returned to Russia while *Leibstandarte* and *Das Reich* were transferred to the West. Over the previous 12 months Hitler had been strengthening the German defenses along the Atlantic Wall. An Allied invasion was expected along the coast of northern France and Belgium, although Holland and Norway were also considered to be vulnerable. Hitler, Rundstedt as C-in-C West, Rommel as C-in-C Army Group B and Schweppenburg as commander of Panzer Group West, were all convinced that the main point of the Allied invasion would be in the area of the Pas de Calais. Other landings would act as diversions to draw in the German panzer reserve. On the eve of the Allied invasion four of the 10 German panzer and panzergrenadier divisions in Belgium and France were Waffen SS. *Leibstandarte* was refitting in Belgium and was well placed to intervene in the Pas de Calais area; *Das Reich* was in southern France at Montauban; *Hitlerju-*

gend at Evreux to the east of Caen; and *Götz von Berlichingen* in the Le Mans area north of the Loire.

None of these divisions was fully up to strength: *Leibstandarte* had two motorized infantry regiments, a panzer regiment and support units, but was not fully re-equipped as a panzer division; *Das Reich* had been reinforced with 9000 replacements, including Alsatians and Luftwaffe personnel. With replacements drawn from 12 nationalities, training proved a challenge to the veteran cadre of *Das Reich*. *Götz von Berlichingen* had only been established in October 1943 and by June 1944 only two-thirds of the division was ready for combat. As late as April one of its antiaircraft batteries relied upon oxen to tow its guns.

Hitlerjugend was considered to be a unique Waffen SS division, one recruited from the racially, ideologically and physically fit youth of Nazi Germany. In 1943 the Hitler Youth leader, Artur Axmann, had suggested to Himmler that Hitler Youth boys born in 1926 should provide the manpower for a complete Waffen SS division. Himmler was enthusiastic and Hitler approved the idea of creating such a new division. *Leibstandarte* provided the cadre of experienced officers and NCOs, including the 35-year-old Fritz Witt as divisional commander, 'Panzer' Meyer and Wilhelm Mohnke to command the panzergrenadier regiments and Max Wünsche to command the panzer regiment. All the 17-year-old Hitler Youth recruits were supposed to be volunteers, but considerable coercion had to be used to get the required numbers. In June 1944 the division was still short of officers and NCOs, but with a divisional strength of 20,500, 177 tanks and 300 armored vehicles, *Hitlerjugend* was a formidable division. On 6 June *Hitlerjugend*

was with the army's *Panzer Lehr* and 21st Panzer Division in Dietrich's I SS Panzer Corps.

As the Allies landed on and behind the Normandy beaches the German defenses were paralyzed. At an operational level it was several hours before senior German officers were convinced that the landings were actually taking place, and once they had been confirmed Hitler was sure they were a diversion to draw in his panzer reserve before the main landings at the Pas de Calais, an opinion he was to hold for several crucial days. Physically, the German defenses were hammered by Allied air attacks, naval bombardments and sabotage by the French Resistance. Taken together, these factors enabled the Allies to build-up their forces relatively unmolested while the Germans only counterattacked after a significant delay.

One of the first Waffen SS units to engage the Allies was *Hitlerjugend*. Motoring to the sound of the guns, Meyer's panzergrenadier regiment moved northwest of Caen to prevent the Canadians from capturing Carpiquet airfield. While *Hitlerjugend* engaged the Canadians other Waffen SS divisions were gradually drawn into the Normandy battle. *Das Reich* began a slow trek north by road and rail distracted by operations against the Resistance. In reprisal for an attack on one of its units, one company killed most of the inhabitants and destroyed the village of Oradour-sur-Glane northwest of Limoges. *Das Reich* was then deployed north of Constances and St Lô and along the Villers-Caen road to contain the Allies by a series of attacks. Units of *Götz von Berlichingen* deployed against the Americans at Carentan.

Hitler began to realize that to push the Allies back into the sea he would have to concentrate sufficient panzer divisions for a major offensive. *Leibstandarte* was ordered to Normandy and Hausser's II SS Panzer Corps from Russia, but it would take Hausser a week before his corps would be in a position to attack. Army divisions and *Hitlerjugend* were tasked with stopping the ponderous advance of Montgomery's forces at Caen. With Allied air superiority and fire support from the sea the Germans were completely vulnerable during the day, and major movement of forces could only be carried out under the cover of darkness. The effectiveness of Allied air and naval bombardments was demonstrated on 10 June when Schweppenburg's Panzer Group headquarters was destroyed by an air attack, and on the 14th when Fritz Witt was killed when *Hitlerjugend*'s divisional head-

TOP LEFT: Captured soldiers of *Götz von Berlichingen* at Courtances, France, July 1944.

ABOVE LEFT: One of the countless casualties of the *Hitlerjugend* in Normandy.

LEFT: Breakfast for Waffen SS troops in Russia 1944.

ABOVE: SS Obersturmführer Michael (in black) Wittmann and his panzer crew in Normandy.

the German panzer reserves leaving the German divisions facing the Americans farther to the west vulnerable to a breakout attack. At the end of June II SS Panzer Corps launched a counterattack to relieve pressure on Dietrich's I SS Panzer Corps. Fierce British resistance and devastating air and sea bombardments forced II SS Panzer Corps to abandon its attack within 24 hours. The British continued to attack along the Caen front and by the second week in July the town was in their hands. German casualties had been heavy. From the 6th *Hitlerjugend* had lost 60 percent of its manpower and half its tanks and armored fighting vehicles. On 18 July the British launched Operation Goodwood from the Orne bridgehead, and although German Army and Waffen SS divisions fought back, they were unable to prevent the loss of the northern slope of the Bourguebus Ridge. On the 25th the US First Army broke through the German defenses at St. Lô and five days later had reached Avranches.

Over the next fortnight, Field Marshal von Kluge, who had succeeded Rundstedt as C-in-C West, attempted to prevent the Americans from achieving a complete breakthrough to the southeast. He began to shuffle panzer divisions, including *Leibstandarte*, westward, but it was too little and too late. The concentric attack of the British Army, north of Falaise, and the US Army south of Argentan, threatened to trap all the German divisions in Normandy. In reality, they were divisions in name only, reduced to kampfgruppen of a few thousand men and a dozen armored vehicles. *Hitlerjugend*, reduced to 15 tanks and 600 men, attempted to hold the Caen-Falaise road. A mixed force of army and Waffen SS units held open a small gap at the eastern end of what became the Falaise-Argentan Pocket, and this enabled about 50 percent of the trapped German forces to escape.

On 21 August, with the Allies pursuing the wreck of the German forces in Normandy toward the Seine, Dietrich assumed command of the German front from the Channel to south of Paris. His command consisted of the broken remnants of once powerful divisions – *Hitlerjugend* had assembled 10 tanks and 300 men. The Germans were unable to hold the line of the Seine and the Allies had pursued them into Belgium and to the Dutch border by 8 September. On 4 September Hitler withdrew the remnants of *Leibstandarte*, *Das Reich* and *Hitlerjugend* to Germany for a complete refit. In the three months following the Normandy invasion German Army and Waffen SS divisions in the West had been routed

quarters came under naval bombardment. Witt was succeeded in command by 'Panzer' Meyer.

Despite the Allied superiority in men and materials, the Germans succeeded in holding the British around Caen. The rough *bocage* of Normandy with its dense hedgerows and sunken lanes favored the defender, while the Tiger and Panther tanks were better than the British tanks. The Germans made effective use of their antitank guns, including the antiaircraft guns in the tank-killer role. In the case of the *Hitlerjugend*, the British and Canadians faced brave and fanatical young soldiers led by experienced and ruthless veterans who had learnt about war in Russia. This fanaticism and ruthlessness resulted in the murder of a number of Canadian prisoners of war.

The attritional battle fought by Montgomery around Caen began to draw in all

and destroyed. Precious little stood between the Allies and Germany. But the Allies, faced with serious logistical problems, and exercising their traditional caution, paused, and allowed the Germans a vital few days to reorganize their defenses.

The campaign in Normandy had tested some of the best Waffen SS divisions. The Waffen SS had provided the backbone of the German battle, and their large and well armed divisions had fought well. Under the leadership of a cadre of battle-hardened officers and NCOs, they were responsible for preventing the Allies achieving a quick, decisive victory in the first few weeks. But the inexperience of many SS soldiers and a casual attitude toward casualties did produce tactical failures and unnecessary losses. Loss of senior German Army officers either through combat or dismissal by Hitler saw both Dietrich and Hausser being given greater responsibility. Following the death of General Dollman, Hausser was appointed to command the Seventh Army. Normandy saw army and Waffen SS units fighting together in tactical and operational units, and the experience of the bitter fighting produced a united camaraderie at the front. The realities of combat in Normandy saw Dietrich and Hausser advising Hitler to discontinue an attack against Avranches during the fighting in the Falaise-Argentan Pocket.

While coping with the crisis in the West Hitler had also to face a series of disasters on the Eastern Front. On 22 June the Soviets had launched a major offensive against Army Group Center, which within six weeks had torn a great gap in the front, destroyed 30 German divisions, and taken the Red Army to the gates of Warsaw. *Totenkopf* and *Wiking*, forming a new IV SS Panzer Corps, counterattacked and held the Soviets in the eastern suburbs of Warsaw until January 1945. The collapse of the Germans and the rapid advance of the Soviets had prompted the Polish Home Army to rise up and attempt to seize Warsaw on 1 August. For the next eight weeks a savage battle was fought for the control of Warsaw between the Polish Home Army and an assortment of SS, police and army units under SS Obergruppenführer Erich von dem Bach-Zelewski, Himmler's expert on anti-partisan warfare. The only Waffen SS units under von dem Bach's command were some Hungarian *Volksdeutsche* from *Maria Theresa* and the infamous Kaminski and Dirlewanger Brigades. Some 150,000 Polish civilians were killed during the fighting, many of them murdered by the SS and police. The Polish Home Army capitulated

at the beginning of October.

As Army Group Center collapsed, the Red Army attacked Army Group North forcing it to retreat, and as the whole of the German Eastern Front came unhinged, Army Group North found itself isolated in Courland by October. The Soviets attacked Army Group North Ukraine, and then in August a new Soviet offensive in the south forced Romania and then Bulgaria to switch from Germany to supporting the Soviet Union. This was followed by a chaotic German retreat from Greece and Yugoslavia in the face of partisan attacks and the Soviet advance. In Yugoslavia, *Prinz Eugen* along with *Skanderbeg* and *Kama*, was swept along in the disaster, while *Florian Geyer* and *Maria Theresa* attempted to stiffen the front in Hungary. *Horst Wessel* was deployed in Hungary but sent a kampfgruppe to assist the German forces defending Galicia and provided units for the German suppression of a revolt in Slovakia.

Germany's allies were beginning to leave the sinking ship, and in October 1944 the Hungarian government of Admiral Horthy attempted to follow the example of Romania. Hitler had been prepared for such treachery and there were sufficient German units in Hungary to prevent a disaster. A special Waffen SS force led by SS Sturmbannführer Otto Skorzeny, carried out a coup d'etat in Budapest and installed a puppet Hungarian government. In September the Finns had withdrawn from the war and forced the German Twentieth Mountain

ABOVE: SS Sturmbannführer Otto Skorzeny.

BELOW: SS Oberführer Dr Oskar Dirlewanger.

Army, including *Nord*, to leave their country. *Reichsführer SS* was deployed in northern Italy attempting to halt the advance of the British Eighth Army. Attacked by Italian partisans, *Reichsführer SS* responded by shooting 2700 civilians and destroying the village of Arno.

Despite this series of major reverses on the Eastern Front, Hitler was determined to use his last remaining reserve of panzer divisions to launch a major offensive in the West and seize the initiative from the Allies. Hitler was convinced that the Waffen SS, his most loyal and committed troops, could achieve victory where the army had failed. His suspicion of the army and been exacerbated by the Bomb Plot on his life on 20 July. Led by army officers based in the headquarters of the *Ersatz* Army in Berlin, the Bomb Plot had nearly killed him. The irony for the Waffen SS given its real role in the Third Reich, was that the suppression of the Bomb Plot was carried out by army units loyal to the Nazi regime. Neither the Waffen SS nor the SS and police were involved, although the latter carried out the subsequent investigation, arrests and interrogations. Apart from its *Ersatz* battalions the field army of the Waffen SS was at the front. Himmler had found that his SS and police units were more than capable of dealing with internal dissent and unrest amongst POWs and forced laborers within Germany and the occupied territories. Following 20 July, Hitler appointed Himmler as C-in-C of the *Ersatz* Army, who in turn appointed Jüttner to deputize for him. With the *Ersatz* Army Himmler had finally laid hands on the manpower and equipment allocation for both the army and the Waffen SS.

On 17 September 1944 Montgomery had attempted to shorten the war in the West by launching Operation Market Garden, whereby British and US airborne forces would seize river crossings in Holland to enable British ground forces to race north and outflank the Westwall. The Waffen SS played a significant role in frustrating the Allied objective. To the east of the Arnhem road bridge, the most northerly of the Allied airborne objectives, II SS Panzer Corps (*Hohenstaufen* and *Frundsberg*) was resting and refitting. After the mauling in Normandy each division consisted of a kampfgruppe of a few thousand men and a few dozen armored vehicles. Bittrich, the corps commander, immediately realized the implication of the air landings, and expeditiously ordered *Hohenstaufen* and *Frundsberg* to deploy and secure the Arnhem bridge, defeat the British airborne forces to the west and hold the road as far south as possible against the British ground attack. The British paratroopers captured the Arnhem road bridge and US airborne troops seized river crossings to the south, but the Allied ground troops could not move quickly enough against determined German resistance and within a week the British airborne force had been withdrawn. With that the objective of Operation Market Garden had been frustrated.

Hitler's operation *Wacht am Rhein*, the offensive in the West, had not been affected by the fighting in Holland. The aim of Hitler's offensive was to penetrate the Ardennes and capture Antwerp. The Allies, thus split and cut off from their supplies, would be encircled and destroyed. It was an attempt to recreate the victory of 1940 when the offensive through the Ardennes had been a brilliant success. But circumstances had changed in four years, and all of them to Germany's disadvantage. Three German armies were to take part in the offensive with Sixth Panzer Army, the most powerful, deployed along the critical northern axis of the attack. Hitler appointed Dietrich to command Sixth Panzer Army, which included in its order of battle *Leibstandarte, Das Reich, Hohenstaufen* and *Hitlerjugend.*

Although the Germans had had two months to prepare for the offensive, the reconstituted divisions bore little resemblance to the veteran formations of a year before. Conscripts, *Volksdeutsche*, Hitler Youth boys, drafts from the army and the Luftwaffe made up the replacements. Veteran Waffen SS officers and NCOs were hard pressed to train and indoctrinate these replacements to the required standards. Many Waffen SS tank drivers were to enter combat in the Ardennes with only a few hours practice. Ideological indoctrination was considered to be of equal importance to training and replacements were given the basic SS creed and accounts of Waffen SS units' bravery and achievements. Dietrich's divisions were re-equipped but never received their full complement of tanks and self-propelled guns.

Sixth Panzer Army had the task of penetrating the US front north of Schnee-Eifel, then race ahead without regard to flank protection, and seize the Meuse crossings on either side of Liège. After several postponements the Ardennes offensive began in bad weather on 16 December. Hitler placed his faith in Dietrich and the Waffen SS to achieve the decisive breakthrough. But even before the offensive Dietrich and his senior Waffen SS officers were very pessimistic, believing that their forces were too weak and poorly trained for the task. Hitler had satisfied Himmler's military ambitions by appointing him as C-in-C of Army Group Rhine, an assortment of army and Waffen SS formations, on 10 December.

Sixth Panzer Army's race to the Meuse was led by a *Leibstandarte* kampfgruppe under SS Obersturmbannführer Joachim Peiper. Confusion and delay at the start of the offensive meant Kampfgruppe Peiper did not begin its advance until late on 16th. Bad roads, blown bridges, fuel shortages and stiffening American resistance meant that Peiper's advance had come to a halt two days later. Other Waffen SS units also ran out of momentum, and Peiper's kampfgruppe became cut off and was eventually forced to abandon its vehicles and filter back through American lines. The axis of the offensive moved away from Sixth Panzer Army to Fifth Panzer Army, but by the New Year the whole offensive had ground to a halt. A new offensive in Alsace, with divisions which included *Nord*, *Frundsberg* and *Götz von Berlichingen*, also failed to achieve a breakthrough.

By 8 January 1945 Hitler had abandoned the Ardennes offensive which had consumed Germany's last significant reserves of fuel and weapons. The Waffen SS had failed him, and he was beginning to doubt their unquestioning obedience and ideological faith. The Allies were now compiling a list of war crimes committed during the war by the army and the Waffen SS. During Peiper's advance in the Ardennes, a group of over 80 American prisoners were shot near the village of Malmèdy and Belgian civilians thought to be partisans were also executed.

During December 1944, while Hitler had been concentrating on the West, the Soviets had not remained inactive. By the 24th Budapest was encircled and a German garrison, mainly consisting of Waffen SS units, including *Florian Geyer* and *Maria Theresa*, was trapped in the city. On 26 December IV SS Panzer Corps, consisting of *Totenkopf* and *Wiking*, was withdrawn from Warsaw

BELOW: Members of the Hitler Youth receive rudimentary military training.

ABOVE: An Hitler Youth boy awarded the iron Cross in March 1945 during the opening stages of the final battle for Berlin.

BELOW RIGHT: Aged recruits to the *Volkssturm* practise with panzerfausts to prevent the final collapse of the Third Reich.

and deployed to Hungary to affect the relief of the garrison. Two attempts in January 1945 failed to relieve Budapest, and a Soviet counteroffensive forced IV SS Panzer Corps onto the defensive north of Lake Balaton. On 12 January the Soviets launched an offensive across the Vistula which, by the end of the month, had taken them to the Oder. This did not affect Hitler's decision to transfer Dietrich's Sixth Panzer Army to Hungary to drive the Soviets away from the oilfields around Lake Balaton and back across the Danube. On 12 February the German garrison surrendered in Budapest. On 5 March Dietrich's Sixth Panzer Army and the IV SS Panzer Corps attacked the Soviets but failed to achieve a breakthrough, and a Soviet offensive on the 16th smashed through the Waffen SS divisions. Despite frantic orders from Hitler to counterattack, Dietrich had no choice but to withdraw into Austria.

Hitler was furious with Dietrich's disobedience and the failure of the Waffen SS. As a mark of their disgrace, Hitler ordered the Waffen SS divisions in Hungary to remove their armbands. To Dietrich and his senior officers who knew about the order, it appeared a complete betrayal of the Waffen SS who had obeyed impossible orders and sacrificed their lives for Hitler. Dietrich was ordered to take part in the defense of Austria, but there was little that the Waffen SS could do to stop the Soviet advance. Waffen SS soldiers fought in the defense of Vienna but the city fell on 13 April. By this stage of the war individual Waffen SS divisions were little more than kampfgruppen, and the *Leibstandarte* was reduced to 16 tanks and 1600 men.

On 23 January 1945 Himmler had been appointed C-in-C Army Group Vistula, an assorted collection of army, Waffen SS, Luftwaffe, *Volkssturm* and Hitler Youth units. Himmler was hopeless as a field commander and by the end of March had resigned with the excuse of poor health. On 16 April the Soviets launched an offensive across the Oder which was to take them to Berlin. The Wehrmacht was brushed aside and the only hope for Hitler trapped in Berlin was a relief force to break the Soviet encirclement. Steiner had been given command of what was loosely described as the Eleventh Panzer Army, a few battalions of military flotsam. Ordered to relieve Berlin, Steiner's army never got across the start line. Hitler was furious and saw another example of Waffen SS disobedience and failure, finally realizing that even the Waffen SS had lost heart and hope.

Then on 28 April he learnt that Himmler had been attempting to negotiate peace terms with the Allies. An enraged Hitler dismissed Himmler from all his offices and appointments and put out an order for his arrest. But it was a futile gesture, and with the last of his army and Waffen SS units defending the area around the Führerbunker, Hitler committed suicide on 30 April. On 7 May the Wehrmacht surrendered unconditionally, and over the next few days Waffen SS troops marched into captivity. Some Waffen SS soldiers committed suicide, many attempted to evade capture and retribution, others like *Das Reich*, *Totenkopf* and *Hitlerjugend* moved off into captivity in military formation, arrogant and defiant to the bitter end.

IN 1946 the SS was indicted as a criminal organization before the International Military Tribunal at Nuremburg and the Waffen SS was included in the charge. In connection with charges of war crimes and crimes against humanity, the tribunal stated that 'units of the Waffen SS were directly involved in the killing of prisoners of war and the atrocities in occupied countries. It supplied personnel for the *Einsatzgruppen* and had command over the concentration camp guards after its absorption of the *Totenkopf* SS, which originally controlled the system.' Several hundred Waffen SS personnel were put on trial before Allied military courts, convicted, and in many cases executed for specific criminal acts. SS Brigadeführer Jürgen Wagner, who commanded *Nederland* at the end of the war, surrendered to the Western Allies, but was extradited to

Canadian prisoners of war in Normandy in 1944. Meyer's death sentence was commuted and he was released in 1954. Seventy-four former members of *Leibstandarte* stood trial before an American military court in 1946 accused of complicity in the massacre of American prisoners of war at Malmèdy in December 1944. Among the accused were Sepp Dietrich, Hermann Priess, Fritz Kraemer and Joachim Peiper. Forty-three of the accused, including Peiper, were condemned to death, 23 to life imprisonment, and the rest to shorter sentences. All of the death sentences were commuted, and the last of the group was released in 1956.

Former members of the Waffen SS defended themselves against the accusation that they had committed war crimes and crimes against humanity. They strongly

CHAPTER 4

A Criminal Fraternity?

Yugoslavia to face prosecution for war crimes, where he was found guilty and executed. SS Obersturmbannführer Fritz Knöchlein of the *Totenkopf* was indicted before a British military court for ordering the shooting of unarmed British prisoners of war at Le Paradis in 1940. He was found guilty and hanged in 1949. SS Gruppenführer Max Simon was tried and condemned to death by a British military court for ordering his division, *Reichsführer SS*, to execute 2700 Italian civilians at Arno in 1944. His sentence was commuted to life imprisonment and he was released in 1954.

A French court tried former members of *Das Reich* for the massacre at Oradour. The accused were junior officers and soldiers, as the officers who ultimately bore responsibility were either dead or could not be extradited. SS Brigadeführer Kurt Meyer was indicted before a Canadian military court and sentenced to death for his role, while serving with *Hitlerjugend*, in the shooting of

objected to the Nuremburg tribunal linking the Waffen SS with the SS itself, and in particular with the concentration and extermination camp systems. Later, when faced with irrefutable evidence establishing a direct link between them and the camps, former Waffen SS soldiers drew a sharp distinction between combat units of the Waffen SS and others in the rear areas or on the home front. 'Panzer' Meyer was to insist that 'the Waffen SS was as much a regular army outfit as any other in the Wehrmacht.' When faced with specific charges indicting Waffen SS men or units for war crimes, former personnel claimed that such acts either had been committed by one or two individuals who could be found in any unit, were the responsibility of others, or had been carried out in the heat of combat.

Postwar Waffen SS objections to any direct link with the SS do not stand up to close scrutiny. Before the war, *Leibstandarte*, SSVT and the SS *Totenkopfverbände*

69

formed part of the SS, were organized and administered by Himmler as Reichsführer SS, and owed direct loyalty and obedience to Hitler. In 1938 Hitler had laid down that the SSVT was a special formation at his personal disposal, part of neither the Wehrmacht nor the police, but in peacetime commanded by Himmler, and regardless of its employment it was a political formation of the Nazi Party. As late as 1942 Hitler was still talking of the Waffen SS as primarily a militarized state protection force whose role was to protect him and the Nazi regime against internal enemies. In wartime that involved service at the front, but also included the extermination of racial and political enemies in the occupied territories.

Himmler played a crucial role in the development of the Waffen SS. As Hitler's 'Ignatius Loyola' Himmler was determined to create a racial and ideological elite based on the SS. The Waffen SS was but one element of this racial elite which would eventually include all parts of the SS and police. Himmler established the ethos of the prewar SSVT, selected its leaders and waged a bureaucratic war with the Wehrmacht on its behalf.

After 1939 Himmler succeeded in persuading Hitler to expand the Waffen SS and nurtured ambitions that one day the Waffen SS would replace the army. The Waffen SS was one part of Himmler's SS and police empire, closely linked with an organization which mushroomed into departments of Security Police and SD, Order Police, concentration and extermination camps, race and resettlement offices, industrial plants and agricultural enterprises. However, Himmler was never really taken that seriously by senior SS officers. Before the war, *Alter Kämpfer* like Dietrich had direct access to Hitler, while former professional soldiers like Hausser tolerated his more absurd ideas. During the war the field commanders of the Waffen SS openly referred to Himmler as '*Reichsheini*' as the gap between the realities of war at the front and the theory of Himmler's world at the rear widened. But on basic questions of racial and political ideology and overall war aims there was little to divide Himmler and the Waffen SS.

In October 1939 the new *Totenkopf* Division had been formed from personnel taken from Eicke's *Totenkopfverbände*. During the first two years of war some 40,000 members of the SS *Totenkopfstandarten*, who were engaged on 'police duties' in the occupied territories, were transferred directly into field units of the Waffen SS. In April 1941 Himmler listed the guard battalions of the

concentration camps as part of the Waffen SS, and these guards wore Waffen SS uniform. During the war there was a considerable exchange of personnel between Waffen SS field units and the concentration camps. Wounded Waffen SS soldiers were sent to guard concentration camps while younger guards were drafted to Waffen SS field units. Four of the most widely publicized and well substantiated atrocities attributed to the Waffen SS – Le Paradis, Oradour, Malmèdy and Arno – were either the work of serving or former members of the *Totenkopf* Division. Le Paradis was the direct responsibility of Fritz Knöchlein, a *Totenkopf* company commander. Oradour was committed by SS soldiers from *Das Reich*, whose commanding general, Heinz

ABOVE: Hitler, with Himmler and Hausser, watches SSVT maneuvers, 1939.

BELOW: Sepp Dietrich in full-dress.

ABOVE: Dutch SS parade before leaving for Russia 1941.

RIGHT: Captured *Totenkopf* concentration camp guards clearing away the dead at Bergen-Belsen, April 1945.

Lammerding, had begun the war serving with *Totenkopf*. Hermann Priess, formerly of the *Totenkopf*, was commanding *Leibstandarte* at the time of Malmèdy and Max Simon was another *Totenkopf* veteran who commanded *Reichsführer SS* which carried out the Arno massacre.

Waffen SS units and personnel were implicated in the racial war and police tasks in the occupied territories. Waffen SS personnel served with the *Einsatzgruppen* in Russia, and *Ersatz* battalions provided personnel for antipartisan warfare and the destruction of the Warsaw Ghetto in 1943. Racial war and special police tasks saw the creation of some particularly odious SS formations. In 1944 Himmler incorporated the Kaminski Brigade into the Waffen SS and placed it under von dem Bach to participate in the suppression of the Warsaw Uprising. Originally formed by a former Soviet engineer, Bronislav Kaminski, as a militia in a semi-autonomous area behind the German lines, the unit consisted of Russians and Ukrainians and gained a reputation for brutality and looting while on antipartisan operations. It withdrew with the retreating

German Army before Himmler put it on Waffen SS books. In Warsaw its brutal murder of thousands of Polish civilians eventually embarrassed Himmler, and Kaminski was killed.

One of the most sinister units which existed on the outer fringes of the Waffen SS was the SS *Sonderkommando* Dirlewanger. In 1940 Gottlob Berger persuaded Himmler to establish a special formation manned by convicted poachers and commanded by his old friend Dr Oskar Dirlewanger. There was one problem about Dirlewanger, although a staunch Nazi and committed anti-semite, he had served time in a concentration camp for sexual offenses against children. But Berger persuaded Himmler that Dirlewanger had other attributes, and in 1942 the SS *Sonderkommando* Dirlewanger was deployed in Russia on antipartisan operations in which its unrestrained brutality and propensity for looting brought forth protests from other SS units. After the supply of German poachers and criminals was exhausted, its personnel was drawn from court-martialed German police and Waffen SS soldiers, *Volksdeutsche*, Russians and

ABOVE: SS security details and Waffen SS men rounding up Jews in 1943 in the Warsaw Ghetto.

BELOW: Walther Krüger, commander of *Das Reich*.

ABOVE: Some of the thousands of Red Army soldiers captured by the Waffen SS.

Ukrainians. Eventually constituted as a brigade, it ended the war as the 36th *Waffen-Grenadierdivision*. Dirlewanger was awarded the Knight's Cross for his part in suppressing the Warsaw Uprising.

Himmler never intended that the Waffen SS should remain isolated from the rest of his SS and police apparatus. Even before the war there was an exchange of personnel, and the SS *Junkerschulen* alumni were not the sole prerogative of the Waffen SS, but could be posted to the Order and Security Police, the concentration camps or the Race and Resettlement Office. Himmler sent a number of SS officers to the Waffen SS who were destined to be higher SS and police leaders in occupied Europe. Before taking up his appointment as higher SS and police leader in southern and then northen Russia, Friedrich Jeckeln served for six weeks in 1940 with *Totenkopf*. Jürgen Stroop, who as SS and police leader in Warsaw in 1943 was to direct the liquidation of the ghetto, served in the summer of 1941 with *Totenkopf* in Russia. By the end of the war personnel exchange between the Waffen SS and the SS and police had become the bureaucratic norm. Many of Himmler's SS and police generals were frustrated soldiers at heart, and the wartime expansion of the Waffen SS gave them the opportunity to exercise military command. SS Obergruppenführer Erich von dem Bach-Zelewski went from commanding Himmler's antipartisan forces on the Eastern Front to commanding an SS Army Corps in 1944.

Two notorious higher SS and police leaders followed a similar pattern. Both SS Obergruppenführer Friedrich Jeckeln and SS Obergruppenführer Friedrich Wilhelm Krüger were to command SS army corps by the end of the war. Perhaps SS Gruppen-führer Heinz Lammerding personified Himmler's fusion between the different parts of his SS empire. Lammerding had started his career as a construction engineer, joined Eicke's prewar *Totenkopf-verbände*, served with the *Totenkopf* until 1943 when he was appointed as chief of staff to von dem Bach, the chief of antipartisan warfare in Russia. In 1944 Lammerding was appointed to command *Das Reich* and ended the war acting as Himmler's chief of staff in Army Group Vistula.

It would be wrong to suggest that only the Waffen SS was responsible for committing atrocities during World War II or that all those who served in the Waffen SS were by definition war criminals. The German Army was fully implicated in the Nazi racial war and in antipartisan operations, including the indiscriminate murder of civilians and Soviet prisoners of war. Army, Waffen SS and SS and police units actively and harmoniously cooperated together in such operations. However, the army and the Waffen SS did make a distinction between the war in the West which, even when it involved the Resistance, was still one between 'civilized peoples,' and the war in the East which was not, as it involved war between German civilization and racial and cultural subhumans. The brutality of the war in the East affected all units who served there and helps to explain some of the savagery demonstrated by Waffen SS units transferred from there in 1944 and stationed in the West. Tens of thousands of Germans, *Volksdeutsche*, drafts sent from the army and the Luftwaffe, West and East Europeans served in the Waffen SS believing they were soldiers not policemen. But the ideology and ethos of the Waffen SS were based on Nazi racism, and a brutality and indifference to human values was carefully inculcated. Simple nationalism and camaraderie existed, as they did in other armies, but were part of a distinct Nazi ideology.

The German Army had an ambivalent attitude toward the Waffen SS. There was always a certain moral repugnance concerning its origins and some of its leading personalities. Before the war there had been a social and professional snobbery about a paramilitary force led by ex-NCOs and amateurs. Undoubtedly, the army always feared that the Waffen SS was a serious rival to its own monopoly position in Nazi Germany. Before and during the war the German Army used every method to restrict the size of the Waffen SS. The expansion of the Waffen SS in 1942-43 owed less to Himmler's intrigues and more to Hitler's criticism of the army's performance compared with that of the Waffen SS. Increasingly, German Army commanders like Guderian and Manstein were impressed by the outstanding fighting spirit and performance of Waffen SS divisions. At the front many of the political, social and ideological distinctions between the army and the Waffen SS became blurred in the cauldron of war, in which they all became 'brothers-in-arms.' By 1945 the Waffen SS had a nominal strength of 830,000 compared with 5,300,000 in the army. The Waffen SS never offered a serious challenge to the army's military position in Nazi Germany; even in the last few months of the war Waffen SS officers only held a handful of corps and army commands, and these were largely illusory.

The self-image and the public image of the Waffen SS has been that of an elite military organization. Hitler and Himmler had established a force which would act as bodyguard, a state protection force and a guard unit for political prisoners. Originally a volunteer force, selected on very strict criteria of race and physical fitness, it was neither part of the army nor the police. The war changed the narrow status defined by Hitler and Himmler with the inclusion of *Totenkopfverbände* and Order Police personnel, followed by *Volksdeutsche*, Germanics, German conscripts and the recruitment of both West and East Europeans. The Waffen SS expanded from 40,000 in December 1939 to 830,000 in April 1945; from three divisions and one regiment in 1939 to 38 divisions and numerous other formations by 1945. During the war at least one-third of Waffen SS manpower was in *Ersatz* battalions and support units. Of the nominal list of 38 divisions, less than half were at full

BELOW: Partisans hanging in Minsk. Draconian measures were the hallmark of SS activities behind the lines in Russia.

casualties was high, reflecting its fanaticism and period of service in direct combat. Thirty-six Waffen SS officers of general rank were killed in action. The casualties among the pre-war SSVT were very heavy, and by 1941 the majority of the prewar SSVT were killed or wounded.

In the early years of the war the Waffen SS had no advantage over the army in terms of weapons and equipment. Indeed, the Waffen SS had to struggle for its share of available weapons and equipment. It was the fighting spirit of the Waffen SS in Russia in 1941-42 which persuaded Hitler to allow expansion and to give it priority in the supply of weapons and equipment. By 1944 Waffen SS panzer and panzergrenadier divisions were larger than the army equivalent, and in some cases had 20 percent more armored vehicles. It was these panzer and panzergrenadier divisions that made the Waffen SS an elite force.

Time and time again Hitler used these divisions to prevent a breakthrough and launch a counterattack, and within months these divisions would be reduced to a combat strength of a few thousand men and a dozen armored vehicles. Around these veterans were grafted volunteer and conscript personnel who frequently neither fulfilled Waffen SS ideological or military requirements. The responsibility for inculcating what had become the Waffen SS ideological and military ethos rested with the veteran survivors. At the divisional level it had been Dietrich, Hausser, Eicke and Steiner, but as they were killed or promoted, it was left to the likes of Fritz Witt, Hermann Priess and 'Panzer' Meyer.

By 1945 the Waffen SS was commanded by young veterans motivated by heroic realism and camaraderie, ruthless and hard toward themselves, their men and the enemy. As one Waffen SS officer was to recall: 'It was those defensive battles of the winter of 1941-42 which I shall always remember, rather than the victorious advance, for the sheer beauty of the fighting. Many of us died horribly, some even as cowards, but for those who lived even for a short period out there, it was well worth all the dreadful suffering and danger. After a time we got to a point when we were not concerned for ourselves, or even Germany, but lived entirely for the next clash, the next engagement with the enemy. There was a tremendous sense of "being," an exhilarating feeling that every nerve in the body was alive to the fight.' The elite Waffen SS were fighters rather than soldiers, and whatever else they were, they were not 'just soldiers like the others.'

ABOVE: SS Standartenführer Fritz Witt.

strength or were ever combat effective, and the rest were regimental size. After 1943 foreigners outnumbered native Germans and 19 of the divisions were manned almost entirely by foreigners. The Waffen SS was never a European Army, but it became an army of Europeans. It was never a force for pan-European idealism and always remained an instrument for the German domination of Europe.

One-third of those who served in the Waffen SS were killed or wounded, and as the majority of those were serving in field formations, the percentage of Waffen SS

List of Waffen SS Divisions

The style of title given to Waffen SS divisions varied, depending on their racial composition. Those composed of German volunteers were styled 'SS Division'; those of *Volksdeutsche* or Germanic volunteers, 'SS Freiwilligen Division'; and those of East Europeans, 'Division der Waffen SS.' By 1944, even 'SS Divisions' had German conscripts and *Volksdeutsche* in their ranks.

1st   SS Panzerdivision *Leibstandarte Adolf Hitler*
2nd SS Panzerdivision *Das Reich*
3rd   SS Panzerdivision *Totenkopf*
4th   SS *Polizei* Panzergrenadierdivision
5th   SS Panzerdivision *Wiking* (Sizeable number of Germanic volunteers)
6th   SS Gebirgsdivision *Nord*
7th   SS Freiwilligen-Gebirgsdivision *Prinz Eugen (Volksdeutsche)*
8th   SS Kavalleriedivision *Florian Geyer* (Sizeable number of *Volksdeutsche*)
9th   SS Panzerdivision *Hohenstaufen*
10th SS Panzerdivision *Frundsberg*
11th SS Freiwilligen-Panzergrenadierdivision *Nordland* (Sizeable number of West Europeans)
12th SS Panzerdivision *Hitlerjugend*
13th Waffen-Gebirgsdivision der SS *Handschar* (kroatische Nr 1) (*Volksdeutsche/*Bosnian Moslems)
14th Waffen-Grenadierdivision der SS (galizische Nr 1) (Ukrainians)
15th Waffen-Grenadierdivision der SS (lettische Nr 1) (Latvians)
16th SS Panzergrenadierdivision *Reichsführer SS* (Sizeable number of *Volksdeutsche*)
17th SS Panzergrenadierdivision *Götz von Berlichingen* (Sizeable number of *Volksdeutsche*)
18th SS Freiwilligen-Panzergrenadierdivision *Horst Wessel (Volksdeutsche)*
19th Waffen-Grenadierdivision der SS (lettische Nr 2) (Latvians)
20th Waffen-Grenadierdivision der SS (estnische Nr 1) (Estonians)
21st Waffen-Gebirgsdivision der SS *Skanderbeg* (albanische Nr 1) (Albanian Moslems never fully formed)
22nd Freiwilligen-Kavalleriedivision der SS *Maria Theresa* (Hungarians/*Volksdeutsche*)
23rd Waffen-Gebirgsdivision der SS *Kama* (kroatische Nr 2) (Bosnian Moslems and never fully formed and dissolved in 1944 and numerical designation given to *Nederland*)

23rd SS Freiwilligen-Panzergrenadierdivision *Nederland* (Dutch and regimental strength)
24th Waffen-Gebirgskarstjägerdivision der SS (Italians/*Volksdeutsche* and regimental strength)
25th Waffen-Grenadierdivision der SS *Hunyadi* (ungarische Nr 1) (Hungarians and reigmental strength)
26th Waffen-Grenadierdivision der SS (ungarische Nr 2) (Hungarians and reigmental strength)
27th SS Freiwilligen-Grenadierdivision *Langemarck* (Flemings and regimental strength)
28th SS Freiwilligen-Grenadierdivision *Wallonien* (Walloons and regimental strength)
29th Waffen Grenadierdivision der SS (russische Nr 1) (Russians and transferred to the Vlasov Army in 1944 and the numerical designation given to italische Nr 1)
29th Waffen Grenadierdivision der SS (italische Nr 1) (Italians and regimental strength)
30th Waffen Grenadierdivision der SS (russische Nr 2) (Russians and regimental strength)
31st SS Freiwilligen-Panzergrenadierdivision *Böhmen-Mähren* (Sizeable number of Foreigners/*Volkdeutsche* formed from Waffen SS schools and training units in Bohemia-Moravia and regimental strength)
32nd SS Panzergrenadierdivision *30 Januar* (Formed from instructors/students from Waffen SS panzer and panzergrenadier schools and regimental strength)
33rd Waffen-Kavalleriedivision der SS (ungarische Nr 3) (Hungarian and destroyed at Budapest in 1945 and numerical designation given to Charlemagne)
33rd Waffen-Grenadierdivision der SS *Charlemagne* (franzosische Nr 1) (French and regimental strength)
34th SS Freiwilligen-Grenadierdivision *Landstorm Nederland* (Dutch and regimental strength)
35th SS Polizei-Grenadierdivision (Order Police mobilized in 1945 and regimental strength)
36th Waffen-Grenadierdivision der SS (Dirlewanger Brigade)
37th SS Freiwilligen-Kavalleriedivision *Lützow* (Largely composed of foreigners and regimental strength)
38th SS Panzergrenadierdivision *Nibelungen* (Formed from officer cadets and instructors from SS *Junkerschule Bad Tölz* and regimental strength)

ABOVE: Part of a poster advertizing a Waffen SS photographic exhibition in Brussels.

RIGHT: Leon Degrelle awards Iron Crosses to survivors of *Wallonien* who had escaped from Cherkassy.

BELOW: SS parade at Nuremberg.

PAGE 78: Hitler touching new SS standards with the 'Blood Flag.'

Knight's Crosses Awarded to Waffen SS Divisions

2nd	SS PzDiv *Das Reich*	72
5th	SS PzDiv *Wiking*	54
1st	SS PzDiv *Leibstandarte*	52
3rd	SS PzDiV *Totenkopf*	46
11th	SS FrwPzGrenDiv *Nordland*	27
8th	SS KavDiv *Florian Geyer*	23
23rd	SS FrwPzGrenDiv *Nederland*	20
4th	SS PolPzGrenDiv	19
12th	SS PzDiv *Hitlerjugend*	15
10th	SS PzDiv *Frundsberg*	13
9th	SS PzDiv *Hohenstaufen*	12
19th	WaffenGrenDiv d.SS (lettische Nr 2)	12
7th	SS FrwGebDiv *Prinz Eugen*	6
6th	SS GebDiv *Nord*	5
18th	SS FrwPzGrenDiv *Horst Wessel*	5
22nd	SS FrwKavDiv	5
13th	WaffenGebDiv d.SS *Handschar*	4
17th	SS PzGrenDiv *Götz von Berlichingen*	4
20th	WaffenGrenDiv d.SS (estnische Nr 1)	4
15th	WaffenGrenDiv d.SS (lettische Nr 1)	3
28th	SS FrwPzGrenDiv *Wallonien*	3
33rd	WaffenGrenDiv d.SS *Charlemagne*	2
14th	WaffenGrenDiv d.SS (galizische Nr 1)	1
16th	SS PzGrenDiv *Reichsführer SS*	1
27th	SS FrwGrenDiv *Langemarck*	1
36th	WaffenGrenDiv d.SS	1
	Total	410

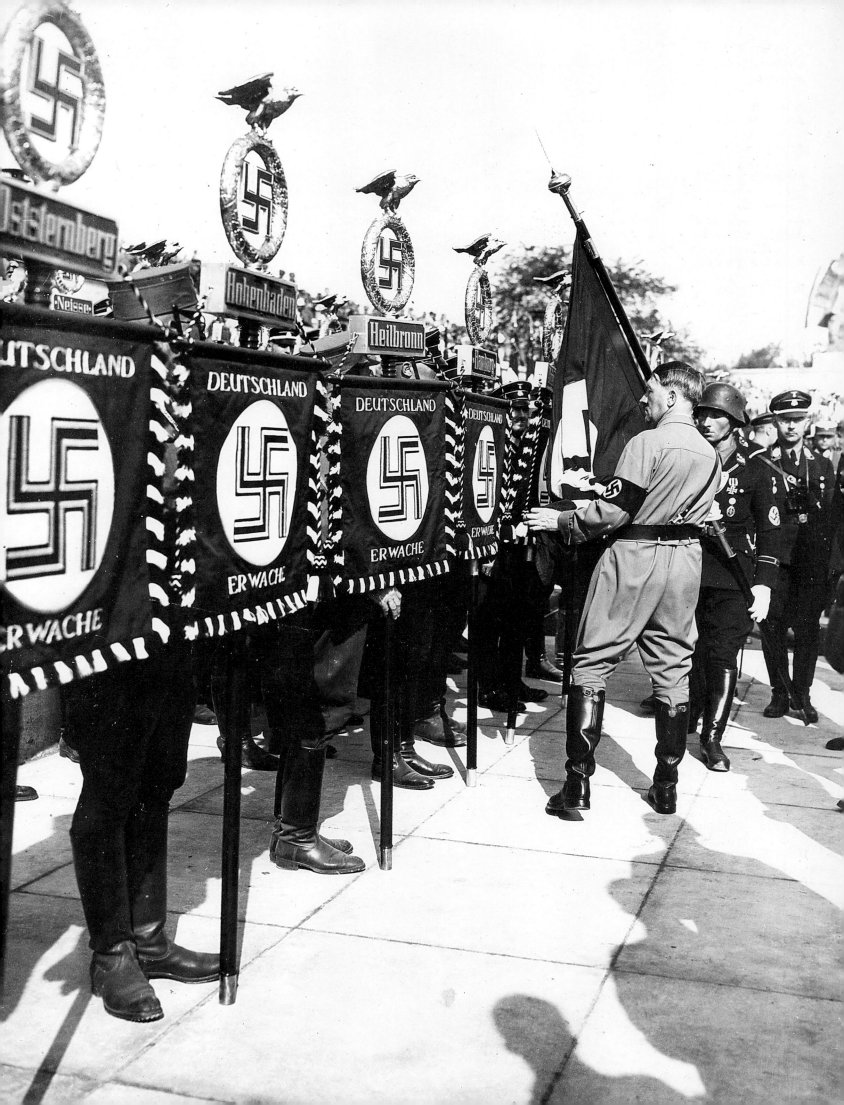

ACKNOWLEDGMENTS

The author and publishers would like to thank Ron Callow for designing this book and Ron Watson for compiling the index. The following agencies and individuals provided photographic material:

Archiv Fur Kunst und Geschichte, pages: 1, 8, 12(top), 15(bottom), 29(bottom), 31, 37(bottom right), 53(top right).
Archiv Gerstenberg, pages: 23(top), 38(top), 70(inset), 72(top), 74, 77(bottom).
Bison Books, pages: 6, 7, 14-15, 42(top), 43(both), 46, 47(bottom), 52(bottom two), 53(bottom two), 56(bottom), 57(top), 61(both).
Bundesarchiv, pages: 4-5, 16(top), 20(top two), 21, 24(both), 28(top), 29(top), 32(middle), 47(top), 57(bottom), 65(left), 66, 67(both), 78.
Collection B L Davis, pages: 17, 19(middle), 20(bottom), 22(both), 25, 26, 29(bottom), 34, 36(left), 37(bottom left), 38(bottom), 39, 40(top), 41, 44(both), 45(top), 48, 49(top), 50, 54(top & bottom), 55(both), 58, 59, 60(both), 62(all three), 64(middle), 72(bottom), 73, 76(top), 77(top).
Robert Hunt Library, pages: 18(top), 32(top & bottom), 33(top), 35, 40(bottom left), 42(bottom), 45(bottom), 63, 65(right), 71(top).
Imperial War Museum, London, pages: 13, 16(bottom), 18(bottom), 33(bottom), 19(top), 36(right), 37(top), 49(bottom), 54(middle), 64(top & bottom), 71(bottom), 75.
Landesbildstelle, Berlin, page: 14(bottom).
MARS, page: 56(top).
Peter Newark's Military Pictures, pages: 2, 10(both), 11, 12(bottom), 19(bottom), 40(bottom right), 52 (top two), 53(top left), 70.
US National Archives, pages: 23(bottom), 68.